SOLID STATE CHEMISTRY

SOLID STATE CHEMISTRY

Arun Rastogi

ANMOL PUBLICATIONS PVT. LTD.
NEW DELHI-110 002 (INDIA)

ANMOL PUBLICATIONS PVT. LTD.

Regd. Office: 4360/4, Ansari Road, Daryaganj,
New Delhi-110002 (India)
Tel.: 23278000, 23261597, 23286875, 23255577
Fax: 91-11-23280289
Email: anmolpub@gmail.com
Visit us at: www.anmolpublications.com

Branch Office: No. 1015, Ist Main Road, BSK IIIrd Stage
IIIrd Phase, IIIrd Block, Bangalore-560 085 (India)
Tel.: 080-41723429 • Fax: 080-26723604
Email: anmolpublicationsbangalore@gmail.com

Solid State Chemistry

First Edition, 2010

ISBN 978-81-261-4470-9

PRINTED IN INDIA

Printed at Mehra Offset Press, Delhi.

Contents

Preface

Solid State Chemistry is a general textbook, designed for those with little background knowledge of the subject, but who wish to learn more about the various segments of solid state theory and technology. Solid-state chemistry is the study of the synthesis, structure, chemical, and physical properties and the application of solid materials.

Covering major developments in the field of solid state chemistry and related areas such as ceramics and amorphous materials, the textbook of Solid State Chemistry features studies of chemical, structural, thermodynamic, electronic, magnetic, and optical properties and processes in solids. The information is presented in a form that can easily be understood and will be useful to readers wishing to build on their own store of knowledge and experience.

Author

Chapter 1

Crystal Structure

INTRODUCTION

A crystal structure is composed of a motif, a set of atoms arranged in a particular way, and a lattice. Motifs are located upon the points of a lattice, which is an array of points repeating periodically in three dimensions. The points can be thought of as forming identical tiny boxes, called unit cells, that fill the space of the lattice. The lengths of the edges of a unit cell and the angles between them are called the lattice parameters. The symmetry properties of the crystal are embodied in its space group. A crystal's structure and symmetry play a role in determining many of its properties, such as cleavage, electronic band structure, and optical properties.

CRYSTALLOGRAPHY

Amorphous solids are homogeneous and isotropic because there is no long range order or periodicity in their internal atomic arrangement. By contrast, the crystalline state is characterised by a regular arrangement of atoms over large distances. Crystals are therefore anisotropic – their properties vary with direction. For example, the interatomic spacing varies with orientation within the crystal, as does the elastic response to an applied stress. Engineering materials are usually aggregates of many crystals of varying sizes and shapes; these polycrystalline materials have properties which depend on the nature of the individual crystals, but also on aggregate properties such as the size and shape distributions of the crystals, and the orientation relationships between the individual crystals.

The randomness in the orientation of the crystals is a measure of texture, which has to be controlled in the manufacture of transformer steels, uranium fuel rods and beverage cans. The crystallography of interfaces connecting adjacent crystals can determine the deformation behaviour of the polycrystalline aggregate; it can also influence the toughness through its effect on the degree of segregation of impurities to such interfaces.

MINERALS

Metals and ceramics are the basic materials of nuclear reactors. The regular arrangement of the atoms in such solids is termed a crystal lattice. Pure solids (whether elemental or binary) need not retain the same crystal structure from low temperature to the melting point; iron and uranium exhibit three crystal structures, each confined to a definite temperature interval, but uranium dioxide has only one. The change from one crystal structure to another is called a phase transformation. Such changes are the solid-solid equivalents of melting and vaporization. The crystal structure can exert a profound effect on material's behaviour in the environment of a nuclear reactor, as seen from the following examples:

- Void swelling can cause unacceptable dimensional changes in structures of fast reactors or fusion reactors (but not generally in light water reactors, except at higher doses and intermediate temperatures in austenitic stainless steels). The crystal structure of the austenitic form of steel is susceptible to this radiation effect but the different crystal structure of the ferritic form is immune until very high doses (>H"50 displacements per atom, dpa).
- Under fast-neutron irradiation, Zircaloy cladding of LWRs undergoes growth in the axial direction. This effect is due to the anisotropy of the crystal structure of zirconium.
- The relation between crystal anisotropy and the

directional dependence of macroscopic properties in zirconium was covered in Sect I.2.

- The utility of uranium metal as a nuclear fuel is diminished because of the volume change accompanying a solid phase transformation that occurs at relatively low temperature.Imperfections in the crystal lattice are even more important than the details of the atomic structure.
- Missing atoms in the atomic lattice called vacancies, or extra atoms in nonregular positions called interstitials, are responsible for the mobility of atoms in the solid
- Many of the mechanical properties of metals are governed by a linear defect in the atomic structure called a dislocation
- All structural alloys and ceramic fuels consist of a large number of small crystals, or grains, joined at their peripheries. The discontinuities in the crystal structure at these internal surfaces are called grain boundaries. Their presence influences a variety of mechanical and transport properties, including cracking in steel and fission gas movement in fuel.

For elemental solids, only 14 distinct structures, or spatial arrangements of atoms, are possible. Each crystal structure is characterized by a unit cell, which is a small group of atoms that contains the unique features of the particular lattice type. A macroscopic crystal is composed of large numbers of unit cells stacked together. The elemental and binary solids important in LWRs are restricted to only a few of the 14 possible lattice types, mainly those with the highest degrees of symmetry and the tightest packing of atoms. With one exception, these unit cells are parallelepipeds (all angles 90o) with side lengths denoted by a, b, and c. The exception is a structure whose unit cell is a hexagonal prism with only two characteristic dimensions. Only the parallelepiped with all sides of equal length (i.e., a cube) is isotropic. All the others are anisotropic, meaning that properties differ in the principal coordinate directions.

THE LATTICE

Crystals have translational symmetry: it is possible to identify a regular set of points, known as the lattice points, each of which has an identical environment. The set of these lattice points constitutes a three dimensional lattice. A unit cell may be defined within this lattice as a space–filling parallelepiped with origin at a lattice point, and with its edges defined by three non-coplanar basis vectors a1, a2 and a3, each of which represents translations between two lattice points. The entire lattice can then be generated by stacking unit cells in three dimensions. Any vector representing a translation between lattice points is called a lattice vector.

The unit cell defined above has lattice points located at its corners. Since these are shared with seven other unit cells, and since each cell has eight corners, there is only one lattice point per unit cell. Such a unit cell is primitive and has the lattice symbol P. Non–primitive unit cells can have two or more lattice points, in which case, the additional lattice points will be located at positions other than the corners of the cell. A cell with lattice points located at the centres of all its faces has the lattice symbol F; such a cell would contain four lattice points.

Not all the faces of the cell need to have face–centering lattice points; when a cell containing two lattice points has the additional lattice point located at the centre of the face defined by a2 and a3, the lattice symbol is A and the cell is said to be A-centred. B-centred and C-centred cells have the additional lattice point located on the face defined by a3 & a1 or a1 & a2 respectively.

A unit cell with two lattice points can alternatively have the additional lattice point at the body-centre of the cell, in which case the lattice symbol is I. The lattice symbol R is for a trigonal cell; the cell is usually defined such that it contains three lattice points. The basis vectors a1, a2 and a3 define the the unit cell; their magnitudes a1, a2 and a3 respectively, are the lattice parameters of the unit cell. The angles a1^a2, a2^a3 and a3 ^a1 are conventionally labelled g,aand b respectively. We see that our initial choice of the basis vectors was arbitrary since there are an infinite number of lattice vectors which could have been used in defining the unit cell. The preferred choice includes small basis vectors which are as equal as possible, provided the shape of the cell reflects the essential symmetry of the lattice.

BRAVAIS LATTICES

The number of ways in which points can be arranged regularly in three dimensions, such that the stacking of unit cells fills space, is not limitless; Bravais showed in 1848 that all possible arrangements can be represented by just fourteen lattices. The fourteen Bravais lattices can be categorised into seven crystal systems (cubic, tetragonal, orthorhombic, trigonal, hexagonal, monoclinic and triclinic); the cubic system contains for example, the cubic-P, cubic-F and cubic-I lattices. Each crystal system can be characterised uniquely by a set of defining symmetry elements, which any crystal within that system must possess as a minimum requirement.

Table. The Crystal Systems

System	Conventional unit cell	Defining symmetry
Triclinic	$a_1 \neq a_2 \neq a_3$, $\alpha \neq \beta \neq \gamma$	monad
Monoclinic	$a_1 \neq a_2 \neq a_3$, $\alpha = \beta$, $\beta > 90°$	1 diad
Orthorhombic	$a_1 \neq a_2 \neq a_3$, $\alpha = \beta = \beta = 90°$	3 diads
Tetragonal	$a_1 - a_2 \neq a_3$, $\alpha - \beta - \beta - 90°$	1 tetrad
Trigonal	$a_1 - a_2 \neq a_3$, $\alpha - \beta - 90°$, $\beta - 120°$	1 triad
Hexagonal	$a_1 = a_2 \neq a_{3'}$ $\alpha = \beta = 90°$, $\beta = 120°$	1 hexad
Cubic	$a_1 = a_2 = a_{3'}$ $\alpha = \beta = \gamma = 90°$	4 triads

The Bravais lattices are each as a projection along the a_3 axis. Projections like these are useful as simple representations of three–dimensional objects. The coordinate of any point with respect to the a_3 axis is represented as a fraction of a_3 along the point of interest; points located at 0*a*3 are unlabelled; translational symmetry requires that for each lattice point located at $0a_3$, there must exist another at $1a_3$.

LINEAR COMBINATIONS

Any vector u can be represented as a linear combination of the basis vectors ai of the unit cell (i = 1, 2, 3):

$$u = u_1a_1 + u_2a_2 + u_3a_3$$

and the scalar quantities u_1, u_2 and u_3 are the components of the vector u with respect to the basis vectors a_1, a_2 and a_3. Once the unit cell is defined, any direction u within the lattice can be identified uniquely by its components [u1 u2 u3], and the components are called the Miller indices of that direction and are by convention enclosed in square brackets.

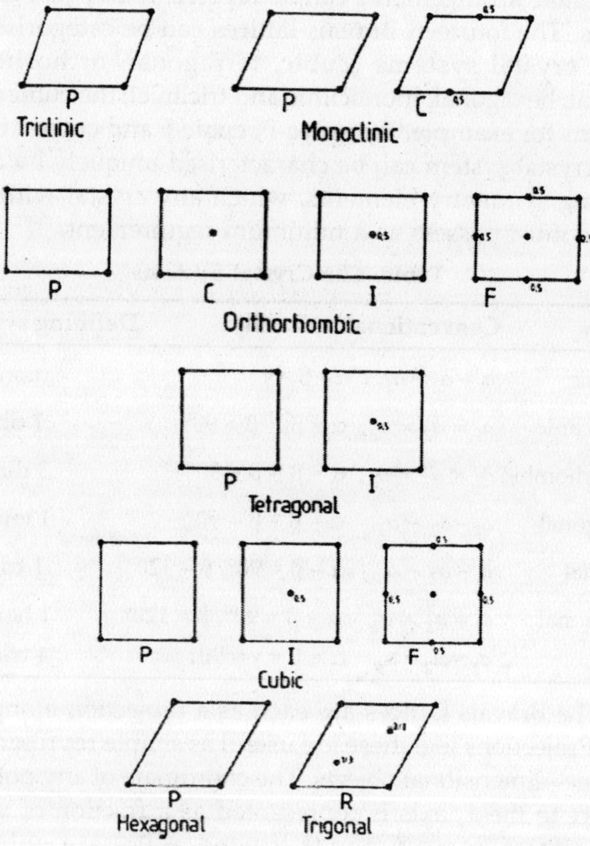

Fig. Projections of the Fourteen Bravais Lattices Along the a_3 Axis

It is sometimes the case that the properties along two or more different directions are identical. These directions are said to be equivalent and the crystal is said to possess symmetry.

For example, the [1 0 0] direction for a cubic lattice is equivalent to the [0 1 0], [0 0 1], [0 1 0], [0 0 1] and [1 0 0] directions; the bar on top of the number implies that the index is negative.

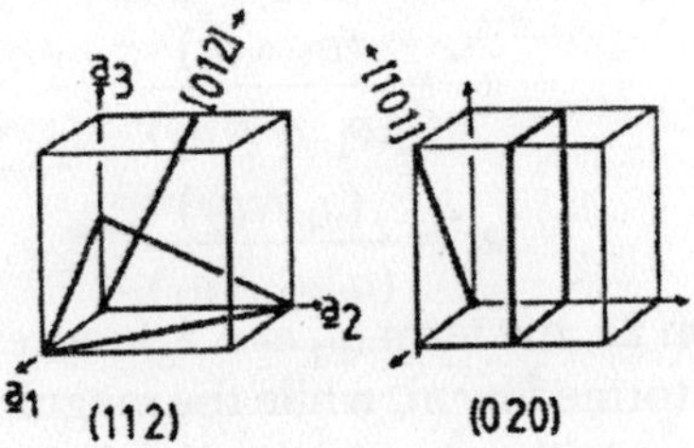

Fig. Miller Indices for Directions and Planes

The indices of directions of the same form are conventionally enclosed in special brackets, e.g. < 1 0 0 >. The number of equivalent directions within the form is called the multiplicity of that direction, which in this case is 6.

PLANES

If a plane intersects the a1, a2 and a3 axes at distances x1, x2 and x3 respectively, relative to the origin, then the Miller indices of that plane are given by (h1 h2 h3) where:

$$h_1 = \phi\frac{a_1}{x_1}, \quad h_2 = \phi\frac{a_2}{x_2}, \quad h_3 = \phi\frac{a_3}{x_3}.$$

ϕ is a scalar which clears the numbers *hi* off fractions or common factors. Note that *xi* are negative when measured in the "*ai* directions.

The intercept of the plane with an axis may occur at 1, in which case the plane is parallel to that axis and the corresponding Miller index will be zero. Miller indices for planes are by convention written using round brackets: (h1 h2 h3) with braces being used to indicate planes of the same form: {h1 h2 h3}.

The Reciprocal Lattice

The reciprocal lattice is a special co–ordinate system. For a lattice represented by basis vectors a1, a2 and a3, the corresponding reciprocal basis vectors are written a_1^*, a_2^* and a_3^*, such that:

$$a_1^* = \frac{(a_2 \wedge a_3)}{(a_1.a_2 \wedge a_3)}$$

$$a_2^* = \frac{(a_3 \wedge a_1)}{(a_1.a_2 \wedge a_3)}$$

$$a_3^* = \frac{(a_1 \wedge a_2)}{(a_1.a_2 \wedge a_3)}$$

In equation 2a, the term $(a_1.a_2 \wedge a_3)$ represents the volume of the unit cell formed by ai, while the magnitude of the vector $(a_2 \wedge$ a3) represents the area of the (1 0 0)A plane. Since $(a_2 \wedge a_3)$ points along the normal to the (1 0 0)A plane, it follows that a_1^* also points along the normal to (1 0 0)A and that its magnitude | | is the reciprocal of the spacing of the (1 0 0)A planes.

The components of any vector referred to the reciprocal basis represent the Miller indices of a plane whose normal is along that vector, with the spacing of the plane given by the inverse of the magnitude of that vector. For example, the reciprocal lattice vector u* = (1 2 3) is normal to planes with Miller indices (1 2 3) and interplanar spacing 1/|u*|.

$a_i \cdot a_j^* = 1$ when $i = j$, and $ai \cdot a_j^* = 0$ when $i \neq j$

or in other words, $a_i \cdot a_j^* = \delta_{ij}$

δ_{ij} is the Kronecker delta, which has a value of unity when $i = j$ and is zero when $i \neq j$

ELECTRONS IN METALS

Metals are electrical conductors because some of the electrons are able to move freely, even though they move in a periodic array of positive metal ions. Electrons are waves, characterised by a wave number k

$$k = \pm\frac{2\pi}{\lambda}$$

where λ is the wavelength. The lattice does not affect the motion of these electrons except at critical values of k, where the lattice planes reflect the electrons†. For a square lattice, these critical values of k are given by

$$k = \frac{n\pi}{a \sin\theta}$$

where a is the lattice parameter and μ is the direction of motion. The component of k along the x axis is $kx = k \sin \mu$ and since μ = 0, it follows that reflection occurs when,

$$kx = \pm \frac{1/4}{a}$$

and similarly when,

$$ky = \pm \frac{1/4}{a}.$$

This can be represented in k–space, which is reciprocal space with a reciprocal lattice parameter of magnitude 2π/a. In Fig. 4, electrons are reflected in the lattice whenever k falls on any point on the square ABCD. This square is known as the first Brillouin zone boundary. Electron energy contours are frequently plotted in k–space. Inside a Brillouin zone, the energy of electrons can increase 'smoothly' with k; there is then a discontinuity (an energy gap) at the zone boundary. This is followed by a second Brillouin zone (n = 2 in equation 3) and so on.

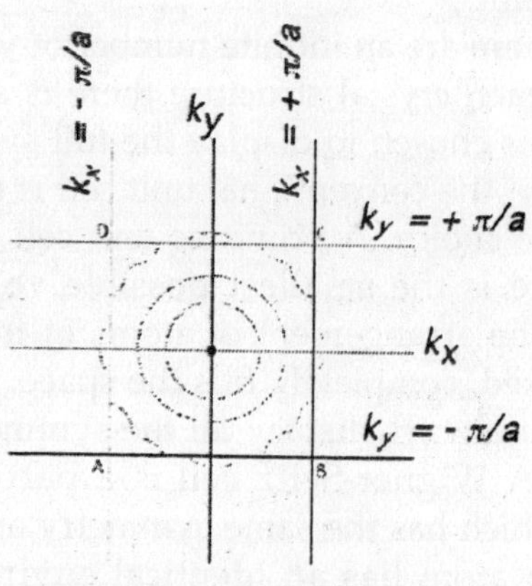

Fig. The Shape of the First Brillouin Zone for the Face–centred Cubic Lattice, Drawn in *k*–space.

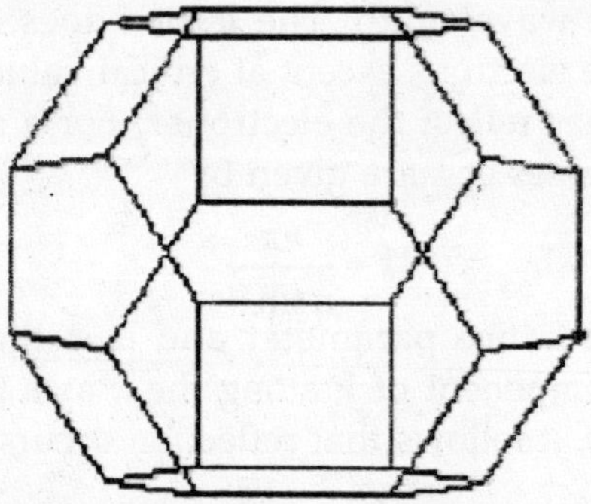

Fig. The Shape of the First Brillouin Zone for a Three–Dimensional Face Centred–cubic Lattice

UNIT CELL

The crystal structure of a material or the arrangement of atoms in a crystal structure can be described in terms of its unit cell. The unit cell is a tiny box containing one or more motifs, a spatial arrangement of atoms. The units cells stacked in three-dimensional space describes the bulk arrangement of atoms of the crystal. The unit cell is given by its lattice parameters, the length of the cell edges and the angles between them, while the positions of the atoms inside the unit cell are described by the set of atomic positions (*xi, yi, zi*) measured from a lattice point.

Although there are an infinite number of ways to specify a unit cell, for each crystal structure there is a conventional unit cell, which is chosen to display the full symmetry of the crystal. However, the conventional unit cell is not always the smallest possible choice. A primitive unit cell of a particular crystal structure is the smallest possible volume one can construct with the arrangement of atoms in the crystal such that, when stacked, completely fills the space. This primitive unit cell will not always display all the symmetries inherent in the crystal. A Wigner-Seitz cell is a particular kind of primitive cell which has the same symmetry as the lattice. In a unit cell each atom has an identical environment when stacked in 3 dimensional space. In a primitive cell, each atom may not have the same environment.

SYMMETRY

Although the properties of a crystal can be anisotropic, there may be different directions along which they are identical. These directions are said to be equivalent and the crystal is said to possess symmetry.

That a particular edge of a cube cannot be distinguished from any other is a measure of its symmetry; an orthorhombic parallelepiped has lower symmetry, since its edges can be distinguished by length.

Some symmetry operations are illustrated in essence, they transform a spatial arrangement into another which is indistinguishable from the original.

The rotation of a cubic lattice through 90° about an axis along the edge of the unit cell is an example of a symmetry operation, since the lattice points of the final and original lattice coincide in space and cannot consequently be distinguished. We have already encountered translational symmetry when defining the lattice; since the environment of each lattice point is identical, translation between lattice points has the effect of shifting the origin.

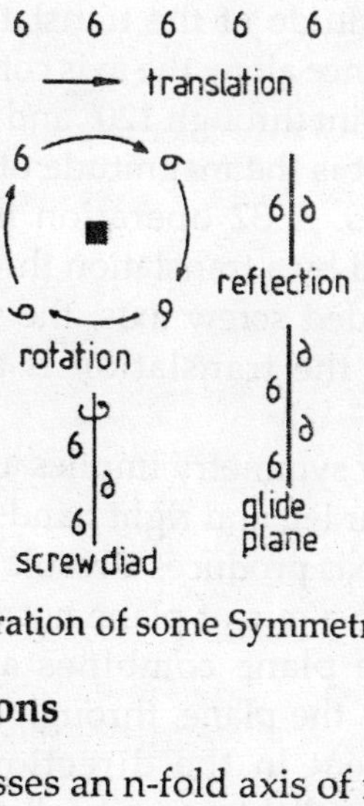

Fig. An Illustration of some Symmetry Operations

Symmetry Operations

An object possesses an n-fold axis of rotational symmetry if it coincides with itself upon rotation about the axis through an angle 360°±/n. The possible angles of rotation, which are consistent with the translational symmetry of the lattice, are 360°, 180°, 120°, 90°and 60° for values of n equal to 1, 2, 3, 4 and 6 respectively. A five–fold axis of rotation does not preserve the translational symmetry of the lattice and is forbidden. A one-fold axis of rotation is called a monad and the terms diad, triad, tetrad and hexad correspond to n = 2, 3, 4 and 6 respectively.

All of the Bravais lattices have a centre of symmetry. An observer at the centre of symmetry sees no difference in arrangement between the directions $[u_1\ u_2\ u_3]$ and $[-u_1\ -u_2\ -u_3]$. The centre of symmetry is such that inversion through that point produces an identical arrangement but in the opposite sense. A rotoinversion axis of symmetry rotates a point through a specified angle and then inverts it through the centre of symmetry such that the arrangements before and after this combined operation are in coincidence. For example, a three-fold inversion axis involves a rotation through 120° combined with an inversion, the axis being labelled $\bar{3}$.

A rotation operation can also be combined with a translation parallel to that axis to generate a screw axis of symmetry. The magnitude of the translation is a fraction of the lattice repeat distance along the axis concerned. A 31 screw axis would rotate a point through 120° and translate it through a distance t/3, where t is the magnitude of the shortest lattice vector along the axis. A 32 operation involves a rotation through 120° followed by a translation through 2t/3 along the axis. For a right–handed screw axis, the sense of rotation is anticlockwise when the translation is along the positive direction of the axis.

A plane of mirror symmetry implies arrangements which are mirror images. Our left and right hands are mirror images. The operation of a 2 axis produces a result which is equivalent to a reflection through a mirror plane normal to that axis. The operation of a glide plane combines a reflection with a translation parallel to the plane, through a distance which is half the lattice repeat in the direction concerned. The translation may be parallel to a unit cell edge, in which case the glide is axial; the term diagonal glide refers to translation along a face or body diagonal of the unit cell. In the latter case, the translation is through a distance which is half the length of the diagonal concerned, except for diamond glide, where it is a quarter of the diagonal length.

CLASSIFICATION OF CRYSTALS BY SYMMETRY

The defining property of a crystal is its inherent symmetry, by which we mean that under certain operations the crystal remains unchanged. For example, rotating the crystal 180 degrees about a certain axis may result in an atomic configuration which is identical to the original configuration. The crystal is then said to have a twofold rotational symmetry about this axis. In addition to rotational symmetries like this, a crystal may have symmetries in the form of mirror planes and translational symmetries, and also the so-called compound symmetries which are a combination of translation and rotation/mirror symmetries. A full classification of a crystal is achieved when all of these inherent symmetries of the crystal are identified.

CRYSTAL SYSTEMS

The crystal systems are a grouping of crystal structures according to the axial system used to describe their lattice. Each crystal system consists of a set of three axes in a particular geometrical arrangement. There are seven unique crystal systems. The simplest and most symmetric, the cubic (or isometric) system, has the symmetry of a cube, that is, it exhibits four threefold rotational axes oriented at 109.5 degrees (the tetrahedral angle) with respect to each other. These threefold axes lie along the body diagonals of the cube. This definition of a cubic is correct, although many textbooks incorrectly state that a cube is defined by three mutually perpendicular axes of equal length – if this were true there would be far more than 14 Bravais lattices. The other six systems, in order of decreasing symmetry, are hexagonal, tetragonal, rhombohedral (also known as trigonal), orthorhombic, monoclinic and triclinic. Some crystallographers consider the hexagonal crystal system not to be its own crystal system, but instead a part of the trigonal crystal system. The crystal system and Bravais lattice of a crystal describe the (purely) translational symmetry of the crystal.

POINT AND SPACE GROUPS

The crystallographic point group or crystal class is the mathematical group comprising the symmetry operations that leave at least one point unmoved and that leave the appearance of the crystal structure unchanged.

These symmetry operations can include reflection, which reflects the structure across a reflection plane, rotation, which rotates the structure a specified portion of a circle about a rotation axis, inversion which changes the sign of the coordinate of each point with respect to a centre of symmetry or inversion point and improper rotation, which consists of a rotation about an axis followed by an inversion.

Rotation axes (proper and improper), reflection planes, and centers of symmetry are collectively called symmetry elements. There are 32 possible crystal classes. Each one can be classified into one of the seven crystal systems.

The space group of the crystal structure is composed of the translational symmetry operations in addition to the operations of the point group.

These include pure translations which move a point along a vector, screw axes, which rotate a point around an axis while translating parallel to the axis, and glide planes, which reflect a point through a plane while translating it parallel to the plane. There are 230 distinct space groups.

PHYSICAL PROPERTIES

IMPURITIES IN CRYSTALS

Real crystals feature defects or irregularities in the ideal arrangements described above and it is these defects that critically determine many of the electrical and mechanical properties of real materials.

When one atom substitutes for one of the principal atomic components within the crystal structure, alteration in the electrical and thermal properties of the material may ensue. Impurities may also manifest as spin impurities in certain materials.

Research on magnetic impurities demonstrates that substantial alteration of certain properties such as specific heat may be affected by small concentrations of an impurity, as for example impurities in semiconducting ferromagnetic alloys may lead to different properties as first predicted in the late 1960s. Dislocations in the crystal lattice allow shear at lower stress than that needed for a perfect crystal structure.

CRYSTAL SYMMETRY PROPERTIES

Twenty of the 32 crystal classes are so-called piezoelectric, and crystals belonging to one of these classes (point groups) display piezoelectricity. All 21 piezoelectric classes lack a centre of symmetry.

Any material develops a dielectric polarization when an electric field is applied, but a substance which has such a natural charge separation even in the absence of a field is called a polar material. Whether or not a material is polar is determined solely by its crystal structure. Only 10 of the 32 point groups are polar. All polar crystals are pyroelectric, so the 10 polar crystal classes are sometimes referred to as the pyroelectric classes.

There are a few crystal structures, notably the perovskite structure, which exhibit ferroelectric behaviour. This is analogous to ferromagnetism, in that, in the absence of an electric field during production, the ferroelectric crystal does not exhibit a polarisation. Upon the application of an electric field of sufficient magnitude, the crystal becomes permanently polarised.

This polarisation can be reversed by a sufficiently large counter-charge, in the same way that a ferromagnet can be reversed.

However, it is important to note that, although they are called ferroelectrics, the effect is due to the crystal structure, not the presence of a ferrous metal. the angle between the normals to the two intersecting faces is called interfacial angle.

Incommensurate crystals have period-varying translational symmetry. The period between nodes of symmetry is constant in most crystals. The distance between nodes in an incommensurate crystal is dependent on the number of nodes between it and the base node.

INTERSTICES

The atoms inside a unit cell do not fill all space. The empty space represents the interstices. It is often the case that these interstices can accommodate small impurity atoms.

As an example, we shall consider the crystal structure of iron which at ambient temperature has the cubic–I lattice with an atom of iron at each lattice point. There are two kinds of interstitial sites capable of accommodating small atoms such as carbon or nitrogen. These are the tetrahedral and octahedral sites as illustrated.

CRYSTALLOGRAPHY AND CRYSTAL DEFECTS

A plane of atoms can glide rigidly over its neighbour in process described as slip with the slip system defined by the plane and the direction of slip. This kind of deformation requires enormous stresses, far greater than those required to actually deform a crystal.

This is because almost all crystals contain defects known as dislocations. A dislocation enables the planes to glide in a piecewise manner rather than the rigid displacement of the entire plane. This greatly reduces the stress required to cause slip. A good analogy to illustrate the role of a dislocation is to imagine the force required to pull an entire carpet along the floor. On the other hand, if a bump is introduced into the carpet, the force needed to move the bump along is much smaller.

The amount and direction of displacement produced by propagating the bump is, in the context of a dislocation, known as its Burgers vector.

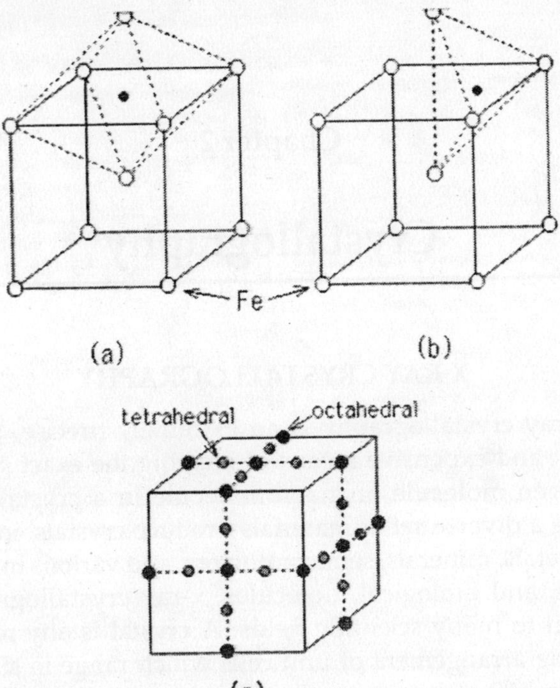

Fig. The Main Interstices in the Body–centred Cubic Structure of Ferrite. (a) An Octahedral Interstice; (b) A Tetrahedral Interstice; (c) Location of both kinds of Interstices.

Crystals may also contain point defects, which are imperfections of occupation. A vacancy is when an atom is missing from a site which should be occupied. An interstitial occurs when an atom is forced into a space within the crystal structure, where atoms are not normally located. Polycrystalline materials contain many crystals; another common term for crystals in such materials is grains. Atoms in the grain boundary between crystals must in general be displaced from positions they would occupy in the undisturbed crystal. Therefore, grain boundaries are defects.

Chapter 2

Crystallography

X RAY CRYSTALLOGRAPHY

X-ray crystallography is an extremely precise, but also difficult and expensive means of imaging the exact structure of a given molecule or macromolecule in a crystal lattice. Because a diverse set of materials produce crystals, including salts, metals, minerals, semiconductors, and various inorganic, organic, and biological molecules, x-ray crystallography is essential to many scientific fields. A crystal is any regularly repeating arrangement of unit cells which range in size from less than 100 atoms — small-molecule crystallography — to tens of thousands — macro-molecular crystallography). X-ray crystallography is famous for being the tool first used to discover the structure of DNA, but it was also used to determine the structure of diamond, table salt, penicillin, numerous proteins, and entire viruses.

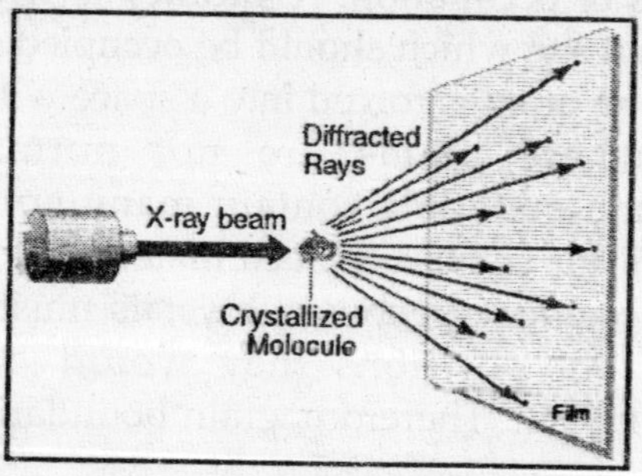

Fig. X-ray Crystallography

In all, over 400,000 structures have been described using x-ray crystallography. These can be found in the Cambridge

Structure Database. To analyse a sample using x-ray crystallography, first one must obtain a high-purity crystal of the material to be studied with a very regular structure. This is often the hardest part as numerous crystals have nanometer-scale defects which make x-ray crystallography difficult.

Next, the sample is subjected to an intense beam of x-rays of a uniform wavelength. These x-rays produce a diffraction pattern as they reflect off the sample. This diffraction pattern is somewhat similar to what is observed when multiple stones are tossed into a pond – where the waves cross are peaks which make up the diffraction pattern. By slowly rotating the crystal, pounding it with x-rays, and meticulously recording the diffraction patterns at each orientation, an electron density map may be derived. This electron density map is then used to formulate a hypothesis about the atomic structure it corresponds to. The diffraction patterns are then analysed in light of the hypothesized structure, and if it looks plausible that the given structure would produce the observed diffraction pattern, a conclusion is drawn.

PRODUCTION OF X-RAYS

X-rays are electromagnetic radiation with wavelengths between about 0.02 Å and 100 Å (1Å = 10-10 meters).

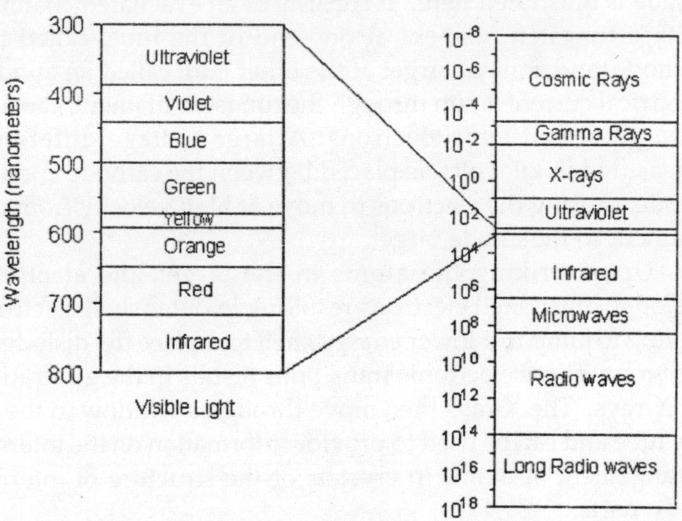

They are part of the electromagnetic spectrum that includes wavelengths of electromagnetic radiation called visible light which our eyes are sensitive to (different wavelengths of visible light appear to us as different colours). Because X-rays have wavelengths similar to the size of atoms, they are useful to explore within crystals.

The energy of X-rays, like all electromagnetic radiation, is inversely proportional to their wavelength as given by the Einstein equation:

E = h = hc

where: E = energy

h = Planck's constant, 6.62517 × 10-27 erg.sec

= frequency

c = velocity of light = 2.99793 × 1010 cm/sec

= wavelength

Thus, since X-rays have a smaller wavelength than visible light, they have higher energy. With their higher energy, X-rays can penetrate matter more easily than can visible light. Their ability to penetrate matter depends on the density of the matter, and thus X-rays provide a powerful tool in medicine for mapping internal structures of the human body.

X-rays are produced in a device called an X-ray tube. Such a tube is illustrated here. It consists of an evacuated chamber with a tungsten filament at one end of the tube, called the cathode, and a metal target at the other end, called an anode. Electrical current is run through the tungsten filament, causing it to glow and emit electrons. A large voltage difference (measured in kilovolts) is placed between the cathode and the anode, causing the electrons to move at high velocity from the filament to the anode target.

Upon striking the atoms in the target, the electrons dislodge inner shell electrons resulting in outer shell electrons having to jump to a lower energy shell to replace the dislodged electrons. These electronic transitions results in the generation of X-rays. The X-rays then move through a window in the X-ray tube and can be used to provide information on the internal arrangement of atoms in crystals or the structure of internal body parts.

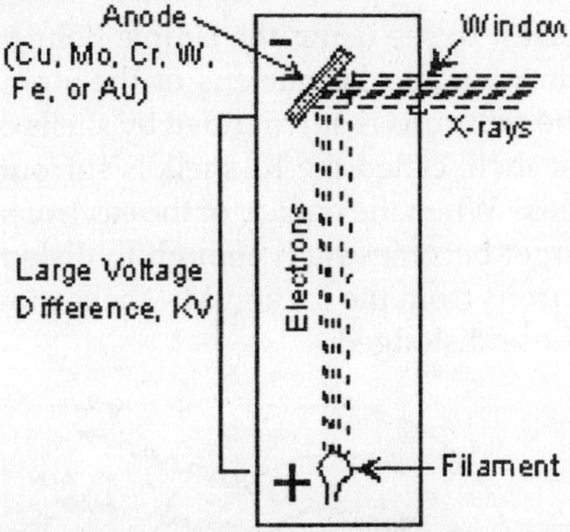

CHARACTERISTIC X-RAY SPECTRA

When the target material of the X-ray tube is bombarded with electrons accelerated from the cathode filament, two types of X-ray spectra are produced. The first is called the continuous spectra. The continuous spectra consists of a range of wavelengths of X-rays with minimum wavelength and intensity (measured in counts per second) dependent on the target material and the voltage across the X-ray tube. The minimum wavelength decreases and the intensity increases as voltage increases. The second type of spectra, called the characteristic spectra, is produced at high voltage as a result of specific electronic transitions that take place within individual atoms of the target material.

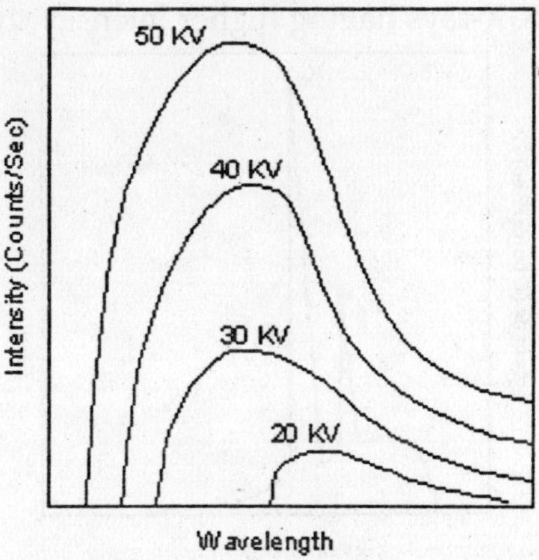

This is easiest to see using the simple Bohr model of the atom. In such a model, the nucleus of the atom containing the protons and neutrons is surrounded by shells of electrons. The innermost shell, called the K- shell, is surrounded by the L- and M - shells. When the energy of the electrons accelerated toward the target becomes high enough to dislodge K- shell electrons, electrons from the L - and M - shells move in to take the place of those dislodged.

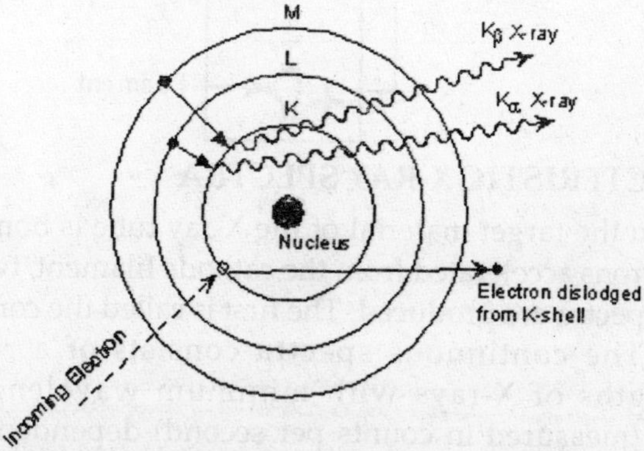

Each of these electronic transitions produces an X-ray with a wavelength that depends on the exact structure of the atom being bombarded. A transition from the L - shell to the K- shell produces a K X-ray, while the transition from an M - shell to the K- shell produces a K X-ray. These characteristic X-rays have a much higher intensity than those produced by the continuous sprectra, with K X-rays having higher intensity than K X-rays.

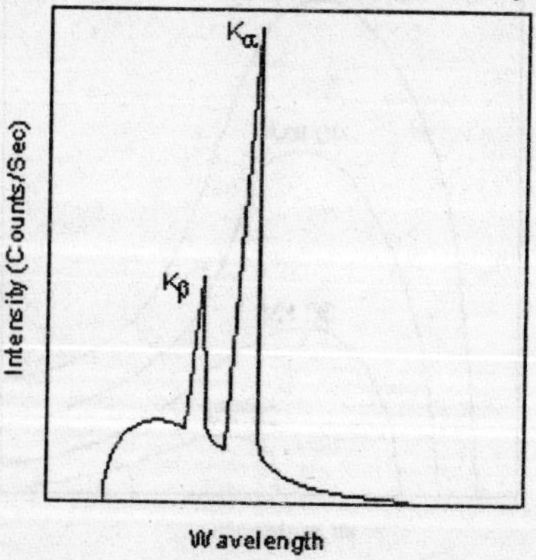

The important point here is that the wavelength of these characteristic x-rays is different for each atom in the periodic table (of course only those elements with higher atomic number have L– and M– shell electrons that can undergo transitions to produce X-rays). A filter is generally used to filter out the lower intensity K X-rays.

For commonly used target materials in X-ray tubes, the X-rays have the following well-known experimentally determined wavelengths:

Element	K Wavelength () Å
Mo	0.7107
Cu	1.5418
Co	1.7902
Fe	1.9373
Cr	2.2909

BRAGG'S LAW

Since a beam of X-rays consists of a bundle of separate waves, the waves can interact with one another. Such interaction is termed interference. If all the waves in the bundle are in phase, that is their crests and troughs occur at exactly the same position (the same as being an integer number of wavelengths out of phase, $n = 1, 2, 3, 4$, etc.), the waves will interfere with one another and their amplitudes will add together to produce a resultant wave that is has a higher amplitude (the sum of all the waves that are in phase.

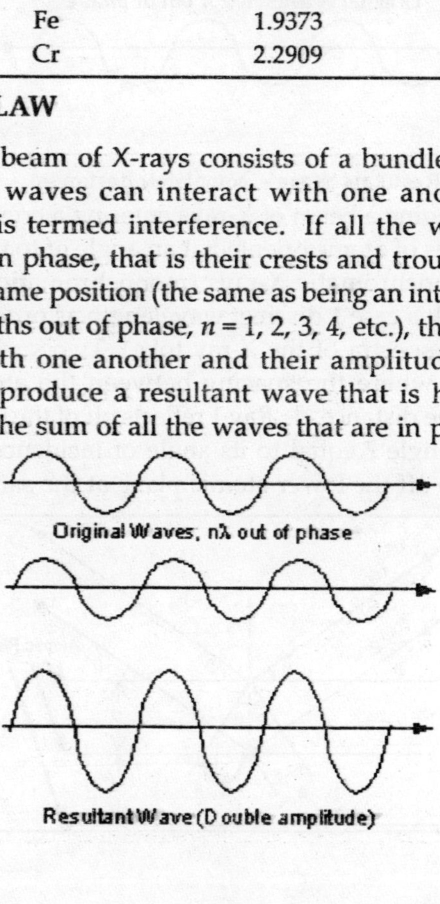

If the waves are out of phase, being off by a non-integer number of wavelengths, then destructive interference will occur and the amplitude of the waves will be reduced. In an extreme case, if the waves are out of phase by a multiple of 1/2 ($n/2$), the resultant wave will have no amplitude and thus be completely destroyed. The atoms in crystals interact with X-ray waves in such a way as to produce interference. The interaction can be thought of as if the atoms in a crystal structure reflect the waves. But, because a crystal structure consists of an orderly arrangement of atoms, the reflections occur from what appears to be planes of atoms.

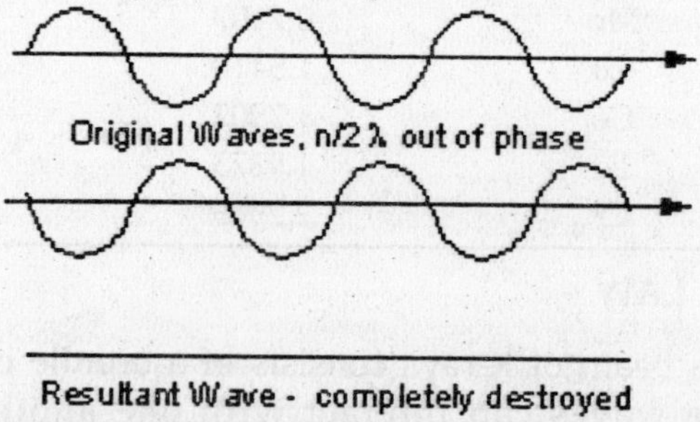

Let's imagine a beam of X-rays entering a crystal with one of these planes of atoms oriented at an angle of to the incoming beam of monochromatic X-rays (monochromatic means one colour, or in this case 1 discreet wavelength as produced by the characteristic spectra of the X-ray tube). Two such X-rays are shown here, where the spacing between the atomic planes occurs over the distance, d. Ray 1 reflects off of the upper atomic plane at an angle ? equal to its angle of incidence. Similarly, Ray 2 reflects off the lower atomic plane at the same angle.

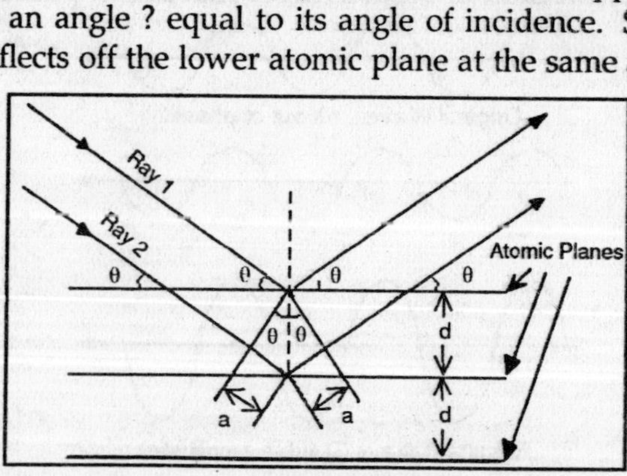

While Ray 2 is in the crystal, however, it travels a distance of 2a farther than Ray 1. If this distance 2a is equal to an integral number of wavelengths (n?), then Rays 1 and 2 will be in phase on their exit from the crystal and constructive interference will occur.

If the distance 2a is not an integral number of wavelengths, then destructive interference will occur and the waves will not be as strong as when they entered the crystal.

Thus, the condition for constructive interference to occur is,

$$n = 2a$$

but, from trigonometry, we can figure out what the distance 2a is in terms of the spacing, d, between the atomic planes.

$$a = d \sin \text{ or } \quad 2a = 2d \sin$$

thus, $n = 2d \sin$

This is known as Bragg's Law for X-ray diffraction.

What it says is that if we know the wavelength ,?, of the X-rays going in to the crystal, and we can measure the angle ? of the diffracted X-rays coming out of the crystal, then we know the spacing (referred to as d-spacing) between the atomic planes.

$$d = \frac{n}{2 \sin}$$

Again it is important to point out that this diffraction will only occur if the rays are in phase when they emerge, and this will only occur at the appropriate value of n (1, 2, 3, etc.) and ?.

In theory, then we could re-orient the crystal so that another atomic plane is exposed and measure the d-spacing between all atomic planes in the crystal, eventually leading us to determine the crystal structure and the size of the unit cell.

X-RAY METHOD

In practice, this would be a time consuming operation to reorient the crystal, measure the angle ?, and determine the d-spacing for all atomic planes.

A faster way is to use a method called the powder method. In this method, a mineral is ground up to a fine powder. In the powder, are thousands of grains that have random orientations. With random orientations we might expect most of the different atomic planes to lie parallel to the surface in some of the grains.

Thus, by scanning through an angle ??of incident X-ray beams form 0 to 90o, we would expect to find all angles where diffraction has occurred, and each of these angles would be associated with a different atomic spacing.

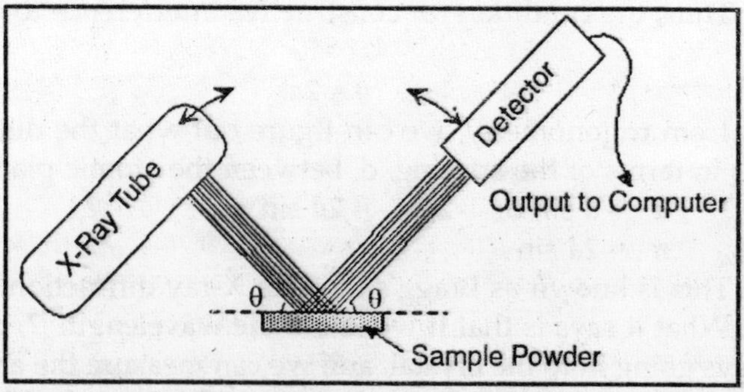

The instrument used to do this is an x-ray powder diffractometer. It consists of an X-ray tube capable of producing a beam of monochromatic X-rays that can be rotated to produce angles from 0 to 90o.

A powdered mineral sample is placed on a sample stage so that it can be irradiated by the X-ray tube. To detect the diffracted X-rays, an electronic detector is placed on the other side of the sample from the X-ray tube, and it too is allowed to rotate to produce angles from 0 to 90o.

The instrument used to rotate both the X-ray tube and the detector is called a goniometer. The goniometer keeps track of the angle and sends this information to a computer, while the detector records the rate of X-rays coming out the other side of the sample (in units of counts/sec) and sends this information to the computer.

After a scan of the sample the X-ray intensity can be plotted against the angle (usually reported as 2 because of the way older diffractometers were made) to produce a chart, like the one shown here. The angle 2 for each diffraction peak can then be converted to d-spacing, using the Bragg equation.

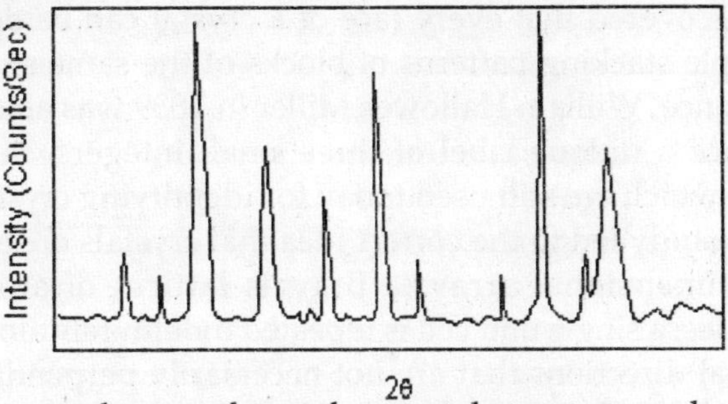

One can then work out the crystal structure and associate each of the diffraction peaks with a different atomic plane in terms of the Miller Index for that plane (hkl). A group known as the Joint Committee on Powder Diffraction Standards (JCPDS) has collected data such as this on thousands of crystalline substances. This data can be obtained as the JCPDS Powder Diffraction File. Since every compound with the same crystal structure will produce an identical powder diffraction pattern, the pattern serves as kind of a "fingerprint" for the substance, and thus comparing an unknown mineral to those in the Powder Diffraction file enables easy identification of the unknown.

CRYSTALS AND X-RAYS

Crystals have long been admired for their regularity and symmetry, but they were not investigated scientifically until the 17th century. Johannes Kepler hypothesized in his work Strena seu de Nive Sexangula that the hexagonal symmetry of snowflake crystals was due to a regular packing of spherical water particles. As shown by X-ray crystallography, the hexagonal symmetry of snowflakes results from the tetrahedral arrangement of hydrogen bonds about each water molecule. The water molecules are arranged simiarly to the silicon atoms in the tridymite polymorph of SiO2. The resulting crystal structure has hexagonal symmetry when viewed along a principal axis.

Crystal symmetry was first investigated experimentally by Nicolas Steno who showed that the angles between the faces are the same in every exemplar of a particular type of crystal who discovered that every face of a crystal can be described by simple stacking patterns of blocks of the same shape and size. Hence, William Hallowes Miller in 1839 was able to give each face a unique label of three small integers, the Miller indices which are still used today for identifying crystal faces. Haüy's study led to the correct idea that crystals are a regular three-dimensional array (a Bravais lattice) of atoms and molecules; a single unit cell is repeated indefinitely along three principal directions that are not necessarily perpendicular.

In the 19th century, a complete catalog of the possible symmetries of a crystal was worked out by Johann Hessel, Auguste Bravais, Yevgraf Fyodorov,, Arthur Schönflies and (belatedly) William Barlow. On the basis of the available data and physical reasoning, Barlow proposed several crystal structures in the 1880s that were validated later by X-ray crystallography; however. the available data were too few in the 1880s to accept his models as conclusive.

X-rays were discovered by Wilhelm Conrad Röntgen in 1895, just as the studies of crystal symmetry were being concluded. Physicists were initially uncertain of the nature of X-rays, although it was soon suspected (correctly) that they were waves of electromagnetic radiation, in other words, another form of light. At that time, the wave model of light — specifically, the Maxwell theory of electromagnetic radiation — was well accepted among scientists, and experiments by Charles Glover Barkla showed that X-rays exhibited phenomena associated with electromagnetic waves, including transverse polarization and spectral lines akin to those observed in the visible wavelengths.

Single-slit experiments in the laboratory of Arnold Sommerfeld suggested the wavelength of X-rays was roughly 1 Angström, one ten millionth of a millimetre. However, X-rays are composed of photons, and thus are not only waves of electromagnetic radiation but also exhibit particle-like properties. The photon concept was introduced by Albert Einstein in 1905, but it was not broadly accepted until 1922, when Arthur Compton confirmed it by the scattering of X-rays from electrons. Therefore, these particle-like properties of X-rays, such as their ionization of gases, caused William Henry Bragg to argue in 1907 that X-rays were not electromagnetic radiation. Nevertheless, Bragg's view was not broadly accepted and the observation of X-ray diffraction in 1912 confirmed for most scientists that X-rays were a form of electromagnetic radiation.

X-RAYS ANALYSIS

The incoming beam (coming from upper left) causes each scatterer to re-radiate a small portion of its energy as a spherical wave. If scatterers are arranged symmetrically with a separation d, these spherical waves will be in synch (add constructively) only in directions where their path-length difference 2d sin θ equals an integer multiple of the wavelength λ. In that case, part of the incoming beam is deflected by an angle 2λ, producing a reflection spot in the diffraction pattern.

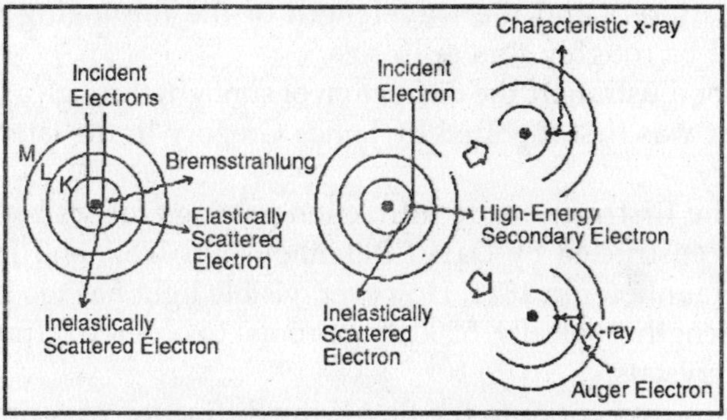

Fig. Spectrum Analysis of X-ray

Crystals are regular arrays of atoms, and X-rays can be considered waves of electromagnetic radiation. Atoms scatter X-ray waves, primarily through the atoms' electrons. Just as an ocean wave striking a lighthouse produces secondary circular waves emanating from the lighthouse, so an X-ray striking an electron produces secondary spherical waves emanating from the electron. This phenomenon is known as scattering, and the electron (or lighthouse) is known as the scatterer. A regular array of scatterers produces a regular array of spherical waves. Although these waves cancel one another out in most directions (destructive interference), they add constructively in a few specific directions, determined by Bragg's law

$$2d \sin \theta = n\lambda$$

where n is any integer. These specific directions appear as spots on the diffraction pattern, often called reflections. Thus, X-ray diffraction results from an electromagnetic wave (the X-ray) impinging on a regular array of scatterers (the repeating arrangement of atoms within the crystal).

X-rays are used to produce the diffraction pattern because their wavelength λ is typically the same order of magnitude (1-100 Ångströms) as the spacing d between planes in the crystal. In principle, any wave impinging on a regular array of scatterers produces diffraction, as predicted first by Francesco Maria Grimaldi in 1665.

To produce significant diffraction, the spacing between the scatterers and the wavelength of the impinging wave should be roughly similar in size.

For illustration, the diffraction of sunlight through a bird's feather was first reported by James Gregory in the later 17th century.

The first man-made diffraction gratings for visible light were constructed by David Rittenhouse in 1787, and Joseph von Fraunhofer in 1821. However, visible light has too long a wavelength (typically, 5500 Ångströms) to observe diffraction from crystals.

However, prior to the first X-ray diffraction experiments, the spacings between lattice planes in a crystal were not known with certainty.

The idea that crystals could be used as a diffraction grating for X-rays arose in 1912 in a conversation between Paul Peter Ewald and Max von Laue in the English Garden in Munich. Ewald had proposed a resonator model of crystals for his thesis, but this model could not be validated using visible light, since the wavelength was much larger than the spacing between the resonators.

Von Laue realized that electromagnetic radiation of a shorter wavelength was needed to observe such small spacings, and suggested that X-rays might have a wavelength comparable to the unit-cell spacing in crystals. Von Laue worked with two technicians, Walter Friedrich and his assistant Paul Knipping, to shine a beam of X-rays through a sphalerite crystal and record its diffraction on a photographic plate.

After being developed, the plate showed a large number of well-defined spots arranged in a pattern of intersecting circles around the spot produced by the central beam.

Von Laue developed a law that connects the scattering angles and the size and orientation of the unit-cell spacings in the crystal, for which he was awarded the Nobel Prize in Physics in 1914.

As described in the mathematical derivation below, the X-ray scattering is determined by the density of electrons within the crystal. Since the energy of an X-ray is much greater than that of an atomic electron, the scattering may be modeled as Thomson scattering, the interaction of an electromagnetic ray with a free electron.

This model is generally adopted to describe the polarization of the scattered radiation. The intensity of Thomson scattering declines as $1/m^2$ with the mass m of the charged particle that is scattering the radiation; hence, the atomic nuclei, which are thousands of times heavier than an electron, contribute negligibly to the scattered X-rays.

GENERATION OF X-RAYS

RONTGEN'S DISCOVERY

X-Rays were discovered by Wilhelm Röntgen in 1895, for which he was awarded the first Nobel Prize in Physics in 1901.

The original records of Röntgen's experiments were lost, but a reconstruction of them in Moore and Reynolds is very interesting. Basically, he discovered that in an experiment with high-energy electrons, another type of radiation was being produced which had the following unique properties:

- Travels in straight lines
- Are exponentially absorbed in matter with the

exponent proportional to the mass of the absorbing material

- Darkens photographic plates
- Make shadows of absorbing material on photo-sensitive paper

The 4th property led directly to the use of x-rays in medicine. In the years following Röntgen's discovery, much was learned about x-rays, and much of the wave vs. particle arguments about radiation which consumed science in the early part of the 20th century were rooted in studies of x-rays.

DISCOVERY OF DIFFRACTION

Studies of x-rays revealed that they could be polarized, but could not be refracted, leading to controversy over whether x-rays were particles or waves. It was understood that if they were waves, the wavelength must be extremely small (10-10 meters or less).

Max von Laue theorized that if x-rays could also be diffracted if the slits were small enough. Since it was understood that molecular spacings in crystalline materials were on the order of a tenth of a nanometer, he devised an experiment in which x-rays were allowed into a lead box containing a crystal, with sensitive film behind the crystal.

When the films were developed there was a large central point from the incident x-rays, but also many smaller points in a regular pattern.

These could only be due to the diffraction of the incident beam and the interference of many beams. By using a crystal as a diffraction grating, von Laue had proved the x-rays were not particles, but waves of light with very small wavelengths. He published his results in 1912.

Lawrence Bragg and his father W.H. Bragg used von Laue's discovery and, for monochromatic radiation, were able to show that diffraction could be treated geometrically like reflection. They derived Bragg's law, which allows diffraction to be treated in simple mathematical terms. The Bragg equation provides a simplified framework for diffraction that works for basic calculations:

$nl \quad = 2d \sin \theta$

Where:

n is an integer

λ is the wavelength of the X-radiation

d is the interplanar spacing in the crystalline material and

λ is the diffraction angle

ELECTROMAGNETIC SPECTRUM

X-Rays are part of the spectrum of electromagnetic radiation occupying the region between the ultraviolet and gamma rays.

SPECTRUM OF ELECTROMAGNETIC RADIATION

The wavelengths of X-radiation commonly used for x-ray diffraction lie between 0.7 and 2.3 **Å**. This is very close to the interplanar spacings of most crystalline materials. The more penetrating radiation used for medical x-rays has a smaller wavelength.

Region	**Wavelength**	**Wavelength**
Frequency	**Energy**	
	(Angstroms)	**(centimeters)**
(Hz)	**(eV)**	
Radio	$> 10^9$	$>10 \quad < 3 \times$
10^9	$< 10^{-5}$	
Microwave	$10^9 - 10^6$	$10 - 0.01 \quad 3 \times 10^9$
$- 3 \times 10^{12}$	$10^{-5} - 0.01$	
Infrared	$106 - 7000$	$0.01 - 7 \times 10^{-5}$
$3 \times 10^{12} - 4.3 \times 10^{14}$	$0.01–2$	
Visible	$7000–4000$	$7 \times 10^{-5} - 4 \times 10^{-5}$
$4.3 \times 10^{14} - 7.5 \times 10^{14}$	$2 - 3$	
Ultraviolet	$4000 - 10$	$4 \times 10^{-5} - 10–^7$
$7.5\times10^{14} - 3 \times 10^{17}$	$3 - 10^3$	
X-rays	$10 - 0.1$	$10^{-7} - 10^{-9} \, 3 \times 10^{17}$
$- 3 \times 10^{19}$	$10^3 - 10^5$	
Gamma Rays	< 0.1	$< 10^{-10} \quad > 3 \times$
10^{19}	$> 10^5$	

A note on units:

an angstrom (Å) is 10^{-10} meters

a nanometer (vm) is 10^{-9} meters

a micrometer (μm) or micron is 10^{-6} meters

a millimeter (mm) is 10^{-3} meters

GENERATIONG OF X-RAYS

To get an accurate picture of the structure of a crystalline material requires X-radiation that is as close to monochromatic as possible.

The function of the x-ray tube and associated electronics is to produce a limited frequency range of high-intensity x-rays. Filters, monochromators, specially tuned detectors and software are then used to further refine the frequency of x-rays used in the analysis.

At left is a schematic of an x-ray tube similar to Cu tubes used in our laboratory.

The arrow from the bottom indicates the direction of electrons generated from a tungsten filament. The filament current is typically set in the range between 25 and 40 ma (milliamps).

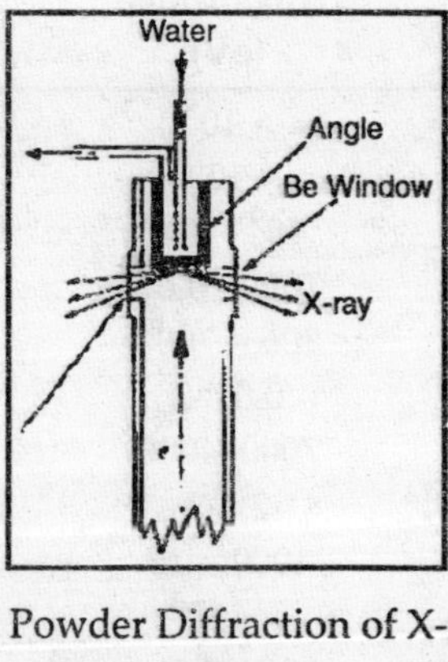

Fig. Powder Diffraction of X-rays

The anode is a pure metal. Cu, Mo, Fe and Cr are in common use in XRD applications. In our lab we use a Cu tube. In one of the XRD systems in Chemistry, a Cr tube is used. Mo is in common use in metallurgical laboratories.

The table below lists the common anodes in use in XRD and their particular advantages and disadvantages. Cooling water (usually kept at about 20 °C) is circulated through the x-ray tube (and sprayed on the back side of the anode) to keep operating temperatures low.

A high potential voltage (typically 30 to 40 KV) is maintained on the anode so that the generated electrons are accelerated and interact with it to generate x-rays.

Electronics are usually designed so that the anode is maintained at ground and a high negative potential placed on the cathode (filament).

Thin metal windows of a light metal (typically Be) that are effectively transparent to x-rays are used to allow x-rays generated to exit in the direction of the specimen and while maintaining the vacuum in the tube. X-ray tubes used for XRD typically have four Be windows to enable multiple ports for output.

CHARACTERISTICS OF X-RAYS

Much of this section is abstracted from Jenkins and Snyder with parts from Moore and Reynolds . X-rays are short-wavelength, high-energy electromagnetic radiation, having the properties of both waves and particles.

They can be described in terms of both photon energy (E) or wavelength, λ (lamda – the distance between peaks) and frequency í (nu – the number of peaks passing a point in a unit of time).

The relationships between these quantities are expressed in the following equations:

$$v = \frac{c}{\lambda}$$

and $E = hv$

where E is the energy of the electron flux in *KeV*

h is Planck's constant (4.135 × 10-15 *eVs*)

c is the speed of light (3 × 1018 Å/*s*)

λ is the wavelength in Å

Substituting the first equation in the second yields:

$$E = \frac{hc}{\lambda}$$

which describes the energy of x-rays in terms of their wavelength. Substituting the values of the constants above in the equation yields the following relationship:

$$E = \frac{12.398}{\lambda}$$

Immediately apparent from this equation is that there is an inverse relationship between the energy and wavelength of x-rays.

Radiation

Successful analysis of powders by x-ray diffraction requires the Bragg diffraction condition to be met, which, in turn, requires high-intensity monochromatic radiation. In the standard powder diffractometer, this radiation is obtained by the generation of very high-intensity "characteristic" x-rays of known wavelength with an x-ray tube.

The characteristic radiation along with the continuous spectrum is then "filtered" by means of an absorptive filter, a crystal monochromator, a wavelength-selective detector or a combination of these devices. Both of the generation of characteristic x-rays and the filtering processes will be addressed in this section.

CHARACTERISTIC X-RAYS

The quantum physics related to the generation of x-rays is, in reality, much more complicated, but this simplified discussion works for our purposes.

If a high-energy electron or photon interacts with electrons in the target anode atoms such that an inner orbital electron is displaced, an electron from a lower energy shell will "jump" to replace the displaced higher-energy electron.

This results in the generation of an x-ray photon of quantum energy equal to that of the transition. The relationship between the energy of this x-ray and its wavelength is described by equation.

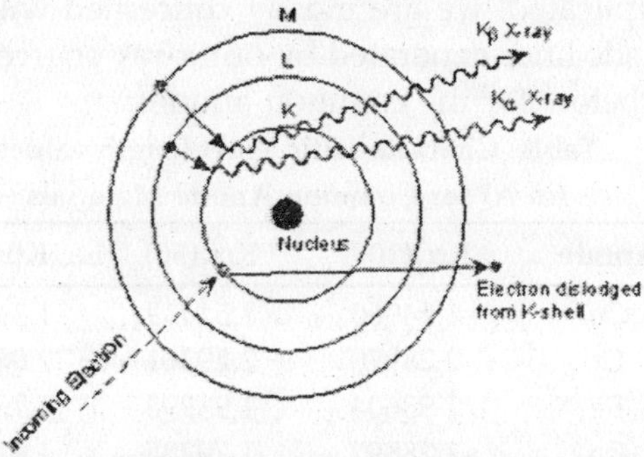

This is shown schematically in the diagram at left. An incoming electron displaces a K-shell electron. If an Lshell electron moves to replace it, a Kα x-ray is produced. If an M-shell electron moves to replace it, a Kα x-ray is produced. The selection rules of quantum transitions in the atom indicate what transitions will occur.

These transitions are shown diagrammatically in the figure above. Two discrete energy "shifts" occur in the L- to K-shell levels transition, yielding $K\alpha_1$ (8.045 keV, 1.5406Å) and $K\alpha_2$ (8.025 keV, 1.5444Å) x-rays. The intensity of the resultant x-rays is related to the number of times the particular transition occurs. The $K\alpha_1$ transition will occur almost exactly twice as frequently as the $K\alpha_2$ transition, thus the resultant x-rays have twice the intensity.

MONOCHROMATIC X-RAYS

There are a number of ways that x-rays "out of the tube" can be modified such that the radiation detected after diffraction is only Kα. Some of these methods include:

- Use of a β filter
- Use of proportional detector and pulse height selection
- Use of a Si(Li) solid-state detector
- Use of a diffracted- or primary-beam monochromator

These methods will be discussed briefly below. Most texts discuss these methods. The treatment by Ron Jenkins in Bish and Post is very concise. The material in this section is abstracted from both sources. Though the x-ray spectrum is more complicated, we are mostly concerned with the strong characteristic lines generated by our x-ray source. These lines are listed below for the common anodes:

Table. Characterisitic Wavelength values (in Å) for Common Anode Materials

Anode	$K\alpha_1$(100)	$K\alpha_2$(50)	$K\beta$(15)
Cu	1.54060	1.54439	1.39222
Cr	2.28970	2.29361	2.08487
Fe	1.93604	1.93998	1.75661
Co	1.78897	1.79285	1.62079
Mo	0.70930	0.71359	0.63229

There is no effective way to remove the $K\alpha_2$ peak from the incident or diffracted beam. At low angles in the diffractogram, the $K\alpha_1$ – $K\alpha_2$ doublet is poorly resolved and will be reflected as a slight asymmetry of the diffraction peak. At higher angles, the doublet will be resolved into two discrete peaks as the angular dispersion of the doublet increases. The known angular dispersion of the doublet and the intensity relationship (the α_2 peak is always about ½ the intensity of the α_2 peak) is used in analytical software to digitally filter the $K\alpha_2$ peak from the resultant diffractogram.

Filters

Filtration is a process by which undesirable characteristic radiation is preferentially absorbed by a filtering material while most of the desired characteristic radiation is passed. Absorption and absorption edges are discussed extensively in Jenkins and Snyder from which most of this discussion is abstracted. All materials absorb x-rays, and several processes operate to produce the mass absorption response of a given material to x-rays. One type of absorption is fundamentally linear and dependent primarily to electron density (mass) of the absorber. Another type is photoelectric absorption, which is related to secondary excitation of orbital electrons in the absorber. This type of absorption is non-linear and varies significantly depending on the wavelength of the x-rays; the point in the wavelength spectrum to which absorption rises gradually and then drops abruptly is called an absorption edge. By selection of a an element for a filter whose absorption edge lies between the characteristic Kβ and Kα wavelength, the Kβ radiation can be greatly attenuated while most of the Kα radiation is passed. A Ni filter is used to filter Cu radiation. Below is the x-ray spectrum for a Copper target xray source overlain by the absorption curve for Ni. When used as a filter, the resultant intensity of the transmitted Cu radiation will be the Cu spectrum minus the Ni absorption curve.

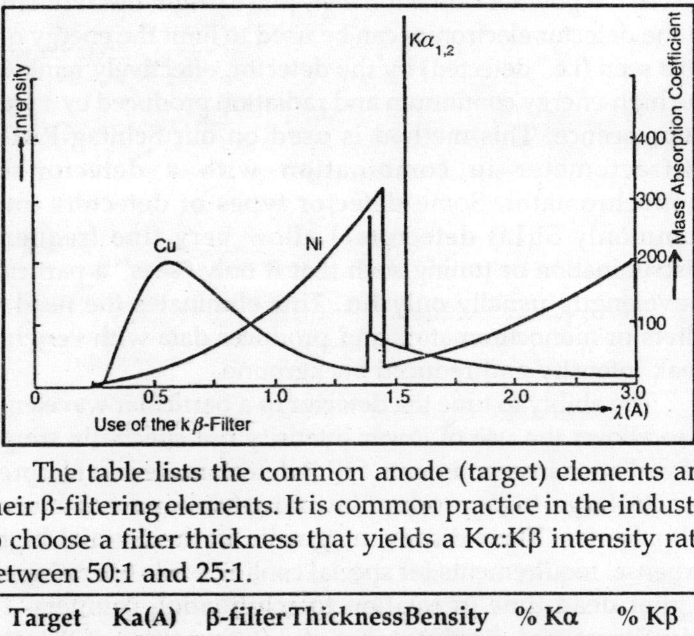

The table lists the common anode (target) elements and their β-filtering elements. It is common practice in the industry to choose a filter thickness that yields a Kα:Kβ intensity ratio between 50:1 and 25:1.

Target	Ka(A)	β-filter	Thickness (μm)	Bensity (g/cc)	% Kα passed	% Kβ passed
Cr	2.291	V	11	6.00	58	3
Fe	1.937	Mn	11	7.43	59	3
Co	1.791	Fe	12	7.87	57	3
Cu	1.542	Ni	15	8.90	52	2
Mo	0.710	Zr	81	6.50	44	1

With a diffractometer, if a filter is used, may be placed on the tube or detector side of the specimen, though the tube-side placement tends to be more common. For Debye-Scherrer powder cameras the filter will be between the source and the camera. As will be noted from the diagram above, while the Kâ radiation is very effectively filtered, a significant amount of the high-energy portion of the continuous spectrum will be passed. In film work this can be a problem and lead to significant fogging of the film; this is one reason that generators used for powder camera work are commonly operated at lower than "optimum" accelerating voltages (i.e., 25 or 30 kV instead of 40 or 45 kV).

In the powder diffractometer, pulse height discrimination in the detector electronics can be used to limit the energy of x-rays seen (i.e., detected) by the detector, effectively removing the high-energy continuum and radiation produced by sample fluorescence. This method is used on our Scintag PAD V diffractometer in combination with a detector-side monochromator. Some detector types of detectors (most commonly Si(Li) detectors1) allow very fine frequency discrimination or tuning such that it only "sees" a particular wavelength, usually only Kα. This eliminates the need for filters or monochromators and produces data with very high peak intensity and reduced background.

The ability to tune the detector to a particular wavelength also allows the use of lower intensity but spectrally simpler Kâ radiation in experiments. Si(Li) detectors are in widespread use in large high-production industrial and commercial laboratories. Their disadvantages lie in their initial higher expense, requirements for special cooling while operating and higher dead time in relation to scintillation counters. The advantages and disadvantages of different types of detectors will be discussed later in this course.

MONOCHROMATORS

The most common way x-rays are filtered on diffractometers, particularly those using scintillation detectors, is by use of a monochromator in either the primary or diffracted beam. These devices make use of the fact that each component wavelength of a polychromatic beam of radiation directed at a single crystal of known orientation and d-spacing will be diffracted at a discrete angle (following the Bragg law). By choosing a crystal with the dspacing to focus only the desired radiation (usually Kα), undesired radiation of all other wavelengths is dispersed and does not enter the detector. Whereas a filter selectively attenuates Kâ, a monochromator selectively passes the desired wavelength and attenuates everything else. High-level bremsstrahlung and other errant radiation from anode contamination (except that in the energy "band" passed by the monochromator) will not get to the detector.

This allows the operation of the x-ray generator optimal accelerating voltages without significantly increasing background "noise" in the detected signal. Possible geometric configurations of monochromators in the beam path are shown in the diagram below and include:

- Diffracted beam parallel,
- Diffracted beam antiparallel,
- Primary beam. In general the diffracted beam parallel configuration is most common; it passes a strong characteristic signal because of minimal path from monochromator to detector, and will remove radiation due to sample fluorescence.

(a) Diffracted-Beam, Parallel

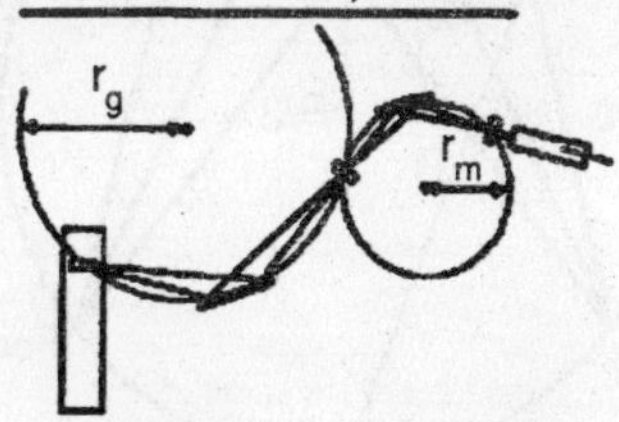

(b) Diffracted-Beam, Antiparallel

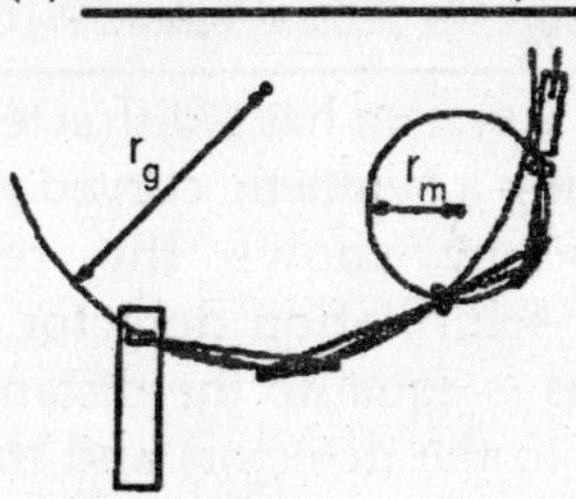

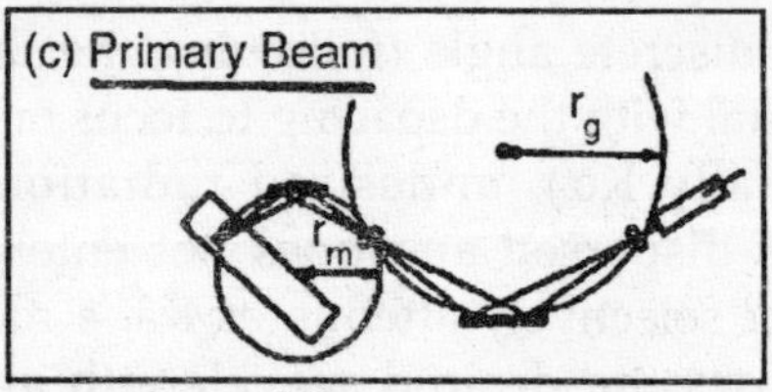

A planar crystal used in a monochromator will tend to diffract the selected wavelength over a very narrow range of incident angles, and will not be very effective in delivering a strong signal to the detector. For this reason, precisely deformed and machined synthetic curved crystals were developed. The pyrolitic graphite curved crystal monochromator is the most widely used crystal monochromator in use. By varying the position of the crystal and the receiving slits within the monochromator assembly, the wavelength passed may be modified. In general, to maximize intensity of the diffracted signal, it is set to pass all Kα radiation, and to not pass Kâ radiation.

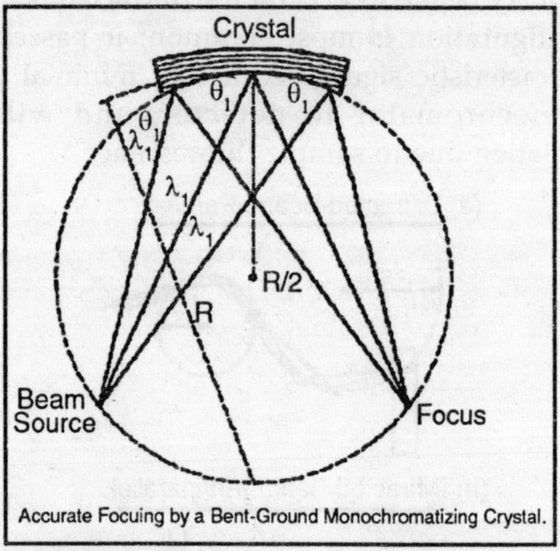

Accurate Focuing by a Bent-Ground Monochromatizing Crystal.

Our Scintag PAD V system has a diffracted beam parallel monochromator utilizing a synthetic curved graphite crystal. The photograph below shows the receiving slits, monochromator and scintillation detector. Note that the monochromator radius is equal to the distance between the receiving slit (closest to the detector) and the centre of the monochromator, and between the centre of the monochromator and the detector.

X-RAY DIFFRACTION

Diffraction can occur when electromagnetic radiation interacts with a periodic structure whose repeat distance is about the same as the wavelength of the radiation. Visible light, for example, can be diffracted by a grating that contains scribed lines spaced only a few thousand angstroms apart, about the wavelength of visible light. X-rays have wavelengths on the order of angstroms, in the range of typical interatomic distances in crystalline solids. Therefore, X-rays can be diffracted from the repeating patterns of atoms that are characteristic of crystalline materials.

Electromagnetic Properties of X-Rays

The role of X-rays in diffraction experiments is based on the electromagnetic properties of this form of radiation. Electromagnetic radiation such as visible light and X-rays can sometimes behave as if the radiation were a beam of particles, while at other times it behaves as if it were a wave. If the energy emitted in the form of photons has a wavelength between 10^{-6} to 10^{-10} cm, then the energy is referred to as X-rays. Electromagnetic radiation can be regarded as a wave moving at the speed of light, c (~3×10^{10} cm/s in a vacuum), and having associated with it a wavelength, λ, and a frequency, ν, such that the relationship $c = \lambda\nu$ is satisfied.

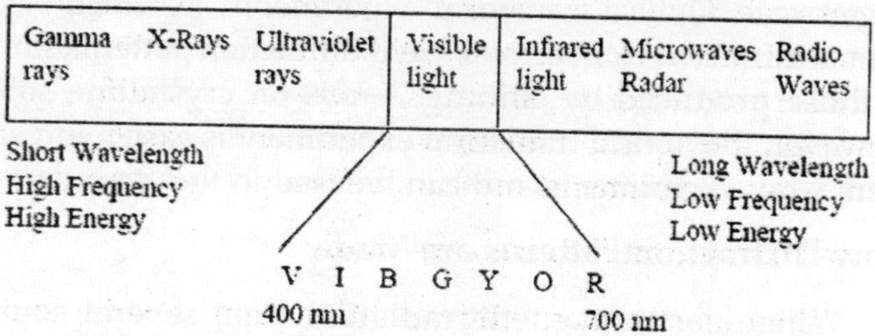

Fig. Electromagnetic Spectrum. The Colours of the Visible Range of the Spectrum are Abbreviated Violet (V), Indigo (I), Blue (B), Green (G), Yellow (Y), Orange (O), and Red (R).

X-Rays and Crystalline Solids

In 1912, Maxwell von Laue recognized that X-rays would be scattered by atoms in a crystalline solid if there is a similarity in spatial scales. If the wavelength and the interatomic distances are roughly the same, diffraction patterns, which reveal the repeating atomic structure, can be formed. A pattern of scattered X-rays (the diffraction pattern) is mathematically related to the structural arrangement of atoms causing the scattering.

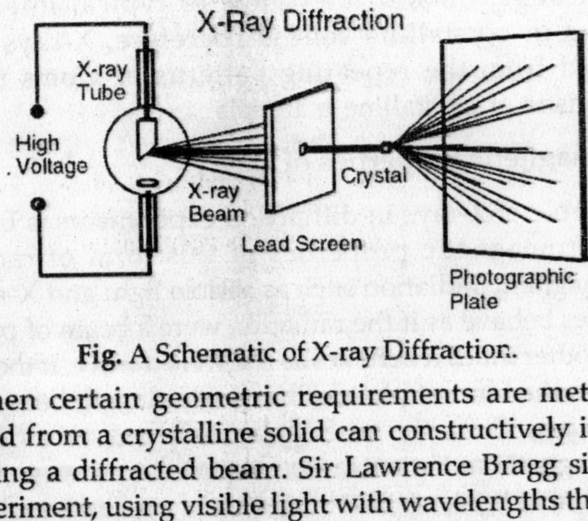

Fig. A Schematic of X-ray Diffraction.

When certain geometric requirements are met, X-rays scattered from a crystalline solid can constructively interfere, producing a diffracted beam. Sir Lawrence Bragg simulated the experiment, using visible light with wavelengths thousands of times larger than those of X-rays. He used tiny arrays of dots and pinholes to mimic atomic arrangements on a much larger scale. Optical transform experiments, in which visible light is diffracted from arrays, yield diffraction patterns similar to those produced by shining X-rays on crystalline solids. However, the optical transform experiment is easier and safer than X-ray experiments and can be used in the classroom.

How Diffraction Patterns are Made

When electromagnetic radiation from several sources overlaps in space simultaneously, either constructive or destructive interference occurs. Constructive interference occurs when the waves are moving in step with one another.

The waves reinforce one another and are said to be in phase. Destructive interference, on the other hand, occurs when the waves are out of phase, with one wave at a maximum amplitude, while the other is at a minimum amplitude. Interference occurs among the waves scattered by the atoms when crystalline solids are exposed to X-rays. The atoms in the crystal scatter the incoming radiation, resulting in diffraction patterns. Destructive interference occurs most often, but in specific directions constructive interference occurs.

Purpose of X-Ray Diffraction

Diffraction data has historically provided information regarding the structures of crystalline solids. Such data can be used to determine molecular structures, ranging from simple to complex, since the relative atomic positions of atoms can be determined. X-ray diffraction provided important evidence and indirect proof of atoms. Diffraction patterns constitute evidence for the periodically repeating arrangement of atoms in crystals. The symmetry of the diffraction patterns corresponds to the symmetry of the atomic packing.

X-ray radiation directed at the solid provides the simplest way to determine the interatomic spacing that exists. The intensity of the diffracted beams also depends on the arrangement and atomic number of the atoms in the repeating motif, called the unit cell. Thus, the intensities of diffracted spots calculated for trial atomic positions can be compared with the experimental diffraction intensities to obtain the positions of the atoms themselves.

From this as well as other indirect methods such as stoichiometric relationships and thermodynamics, evidence of atoms was obtained. However, a direct way to image atoms on the surfaces of materials now exists. Developed in the mid-1980's, the scanning tunneling microscope (STM) permits direct imaging of atoms.

Scanning Probe Microscopy (SPM)

Scanning Probe Microscopy (SPM) includes Scanning Tunneling Microscopy (STM), Atomic Force Microscopy

(AFM), and a variety of related experimental techniques. These are experimental methods that are used to image both organic and inorganic surfaces with (near) atomic resolution. In a scanning tunneling microscope a sharp metal tip, terminating ideally in a single atom, is positioned over an electrically conducting substrate, and a small potential difference is applied between them. The gap between the tip and the substrate surface is made large enough that electrical conduction cannot occur; yet, it is small enough to let electrons tunnel (a quantum mechanical phenomenon) between the tip and the surface. Tunneling probability decays exponentially with increasing tip-to-surface separation. Thus, the spatial arrangement of atoms on the surface is determined by the variation in tunneling current sensed by the probe tip as it moves in atomic-scale increments across the surface, a process called rastering. Scanning is more commonly done by adjusting the tip-to-surface separation so as to maintain a constant tunneling current, thereby preventing the tip from crashing into the surface. In either mode of operation a "map" of the sample surface with atomic resolution results.

Atomic Force Microscopy (AFM)

In an atomic force microscope the surface topography is mapped by measuring the mechanical force between tip and surface rather than the electrical current flowing between them as the STM does. Since force is used to create the images rather than the electrical current, the AFM can be used to image both conducting and non-conducting substrates. To measure the interatomic force, the tip of the AFM is mounted on the end of a small cantilever. As the interatomic force varies, the deflection of the lever can be sensed by bouncing a laser beam off the back of the lever and measuring displacements with a pair of photosensors.

SCANNING TUNNELING MICROSCOPE (STM)

Gert Binnig and Heinrich Rohrer were awarded the Nobel Prize in Physics, in 1986, for the development of the scanning tunneling microscope. They were also jointly honored with

Ernst Ruska for their work on the development of electron microscopy.

To gain a better understanding of how the scanning tunneling microscope works, the behaviour of electrons in metals and other electrically conducting material needs to be considered. Electrostatic forces acting between the electrons and the nuclei of atoms hold the atoms of a metal together. Core electrons are bound tightly to individual nuclei. However, the valence electrons that are farthest from the nuclei feel a relatively weak electrostatic attraction and are free to move about in the space between the nuclei. Since these electrons carry or conduct the electric current, they are referred to as conduction electrons.

The large numbers of valence electron orbitals overlap and provide a continuous distribution of states available to the conduction electrons, called a band, that extends over the entire solid. Each orbital can be occupied by a pair of electrons with opposite spin, and they are filled in order from lowest to highest orbital in energy. The Fermi energy (EF) is the energy of the most weakly bound electrons. The electrons at the Fermi energy are held in the metal by an energy barrier. Classically, these electrons can never leave the metal unless they are given enough energy to go over this potential barrier. Quantum mechanically, however, electrons near the Fermi energy can tunnel through the potential barrier.

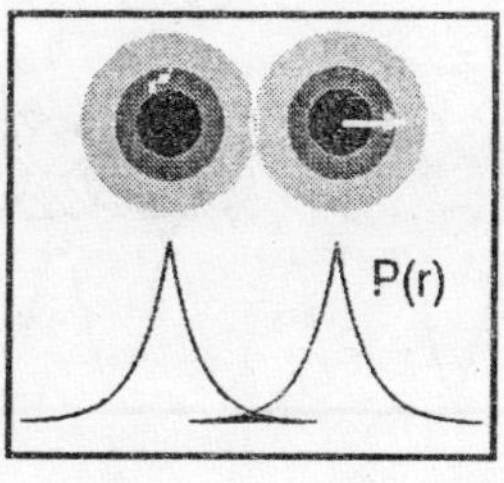

Fig. Two Atoms with their Electron Probability Clouds Slightly Overlapping.

Tunneling

The quantum mechanical phenomenon called tunneling

is possible when the tip is only within a few angstroms (10-8 cm) of the surface.

Tunneling is the term used to describe the movement of an electron through a classical barrier, which is possible only due to its wave nature and hence impossible in classical physics. To understand this better, consider only one atom. The electrons surrounding the atomic nucleus are not confined to a hard shell but are within a varying probability distribution.

This causes the edges of the atom to be indistinct. When the quantum mechanical equations describing the probability of the electron locations are solved, it is found that the electron spends most of its time near the nucleus, and the probability distribution falls off exponentially as the distance from the nucleus increases.

Because the electron probability distribution falls off so rapidly with distance from the nucleus, this tunneling current provides a very sensitive probe of interatomic separation. If two atoms are within angstroms of each other, an electron from one atom can move through the region of overlapping electron density to become part of the other atom's electron.

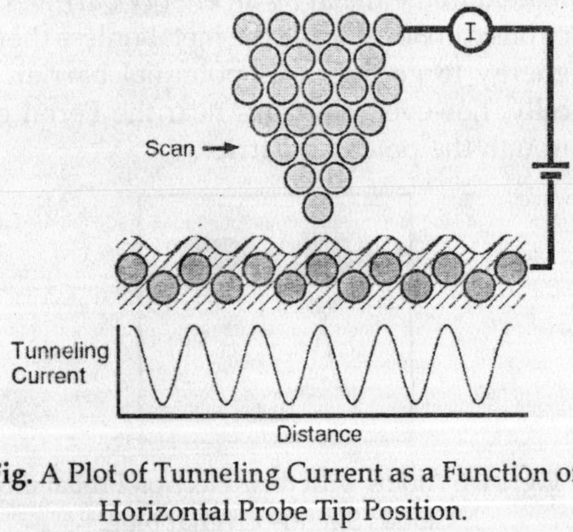

Fig. A Plot of Tunneling Current as a Function of Horizontal Probe Tip Position.

CHALLENGES FOR STM

Theoretically, STM can be used to image individual atoms

on the surface; in practice, however, three challenges arise. The first challenge, vibrations, are important because the separation between the sample and probe is so small. Since the tip is only a few angstroms from the surface, it is easy to crash it into the sample unless the substrate is smooth on the atomic scale.

For such a small separation, any minor perturbation such as vibrations set up by a sneeze or motion in the room can jam the probe into the sample and ruin the experiment. As a result, careful engineering is necessary to make the instrument rigid and to isolate it from external disturbances.

Another problem, probe sharpness, determines how small a structure can be imaged on the surface. Electrochemical etching can be used to sharpen the end of a metal wire to a radius of about 1000 nm.

A probe with such a large surface area would allow tunneling to occur over a large region of the sample surface. In order to detect individual atoms, the probe tip must be comparable in size to an atom.

Thus, the probe tip must ideally consist of a single atom. The final problem is that of position control. In order to move the probe with controllable displacements of 0.1 nm (1 angstrom) or less, a special type of piezoelectric ceramic material is used. This material expands and contracts on a scale of angstroms when appropriate external voltages are applied to pairs of electrodes on its opposite faces.

The STM Tip

The tip is prepared so that it terminates in a single atom. The tip is usually composed of tungsten or platinum. If the experiment is performed in a vacuum, tungsten is the preferred material because it is relatively easy to prepare a single-atom-terminated tip.

If the STM experiment is to be performed in a liquid or in air, tungsten reacts too quickly. Therefore, Pt or Pt-Ir alloys are preferred even though it is more difficult to prepare tips with these materials, and they generally are not as atomically sharp.

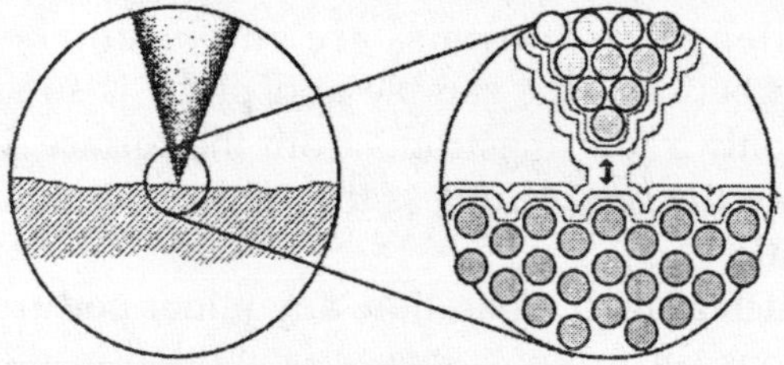

Fig. atomically sharp tip near a surface

USES AND CAPABILITIES OF STM

The STM has many uses. It is used in fundamental studies of the physics of atoms at surfaces. STMs can be constructed to be compatible with high-vacuum conditions, which are used to study the properties of atomically "clean" surfaces or surfaces that have been modified in some controlled way. The STM can also be used to study electrode surfaces immersed in liquid electrolytes.

In addition to these scientific applications, the STM has a wide range of potentially practical applications. The STM can image structures ranging from DNA in a biological environment to the surface of an operating battery electrode. The application of the STM to biological molecules has been proposed as a method of gene sequencing. Research is currently being done to demonstrate the ability to write with atomic resolution. Features a few nanometers wide have been written by using the probe to scratch or dent the surface directly or by using the tunneling current to locally heat the surface of a substance. The probe has even been used to move individual atoms so as to form a word.

Scanning tunneling microscopy is a practical demonstration of quantum mechanics. Scanning probe microscopy techniques may be used to create atomic-scale devices and new structures. For example the STM has been used to prepare a "nanobattery," which consists of two copper pillars and two silver pillars that are placed sequentially on a graphite surface by electrochemical reduction of solutions of copper sulfate and silver fluoride.

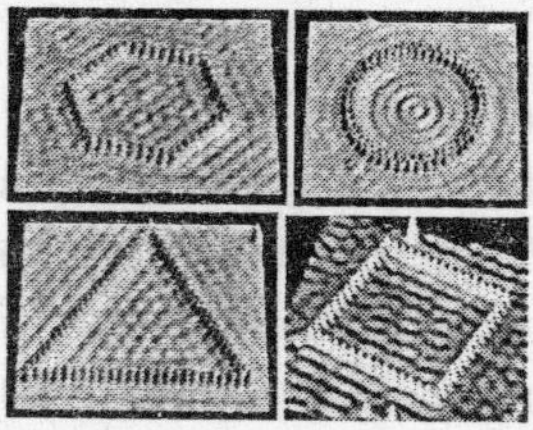

Fig. Scanning Tunneling Microscopy

SINGLE-CRYSTAL X-RAY DIFFRACTION

The oldest and most precise method of X-ray crystallography is single-crystal X-ray diffraction, in which a beam of X-rays strikes a single crystal, producing scattered beams. When they land on a piece of film or other detector, these beams make a diffraction pattern of spots; the strengths and angles of these beams are recorded as the crystal is gradually rotated. Each spot is called a reflection, since it corresponds to the reflection of the X-rays from one set of evenly spaced planes within the crystal. For single crystals of sufficient purity and regularity, X-ray diffraction data can determine the mean chemical bond lengths and angles to within a few thousandths of an Ångström and to within a few tenths of a degree, respectively. The atoms in a crystal are also not static, but oscillate about their mean positions, usually by less than a few tenths of an Ångström. X-ray crystallography allows the size of these oscillations to be measured quantitatively.

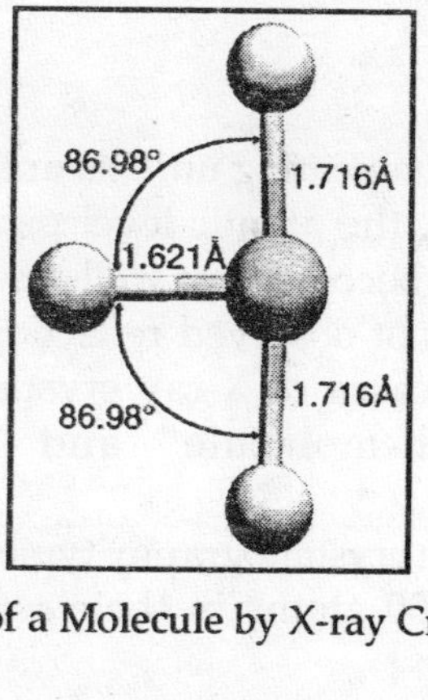

Fig. Structure of a Molecule by X-ray Crystallography

Procedure

The technique of single-crystal X-ray crystallography has three basic steps. The first — and often most difficult — step is to obtain an adequate crystal of the material under study. The crystal should be sufficiently large (typically larger than 100 micrometres in all dimensions), pure in composition and regular in structure, with no significant internal imperfections such as cracks or twinning.

A small or irregular crystal will give fewer and less reliable data, from which it may be impossible to determine the atomic arrangement.

In the second step, the crystal is placed in an intense beam of X-rays, usually of a single wavelength (monochromatic X-rays), producing the regular pattern of reflections. As the crystal is gradually rotated, previous reflections disappear and new ones appear; the intensity of every spot is recorded at every orientation of the crystal.

Multiple data sets may have to be collected, with each set covering slightly more than half a full rotation of the crystal and typically containing tens of thousands of reflection intensities.

In the third step, these data are combined computationally with complementary chemical information to produce and refine a model of the arrangement of atoms within the crystal. The final, refined model of the atomic arrangement — now called a crystal structure — is usually stored in a public database.

Limitations

As the crystal's repeating unit, its unit cell, becomes larger and more complex, the atomic-level picture provided by X-ray crystallography becomes less well-resolved (more "fuzzy") for a given number of observed reflections.

Two limiting cases of X-ray crystallography are often discerned, "small-molecule" and "macromolecular" crystallography.

Small-molecule crystallography typically involves crystals with fewer than 100 atoms in their asymmetric unit; such crystal structures are usually so well resolved that its atoms can be discerned as isolated "blobs" of electron density. By contrast, macromolecular crystallography often involves tens of thousands of atoms in the unit cell. Such crystal structures are generally less well-resolved (more "smeared out"); the atoms and chemical bonds appear as tubes of electron density, rather than as isolated atoms.

In general, small molecules are also easier to crystallize than macromolecules; however, X-ray crystallography has proven possible even for viruses with hundreds of thousands of atoms.

CRYSTALLIZATION

Although crystallography can be used to characterize the disorder in an impure or irregular crystal, crystallography generally requires a pure crystal of high regularity to solve for the structure of a complicated arrangement of atoms.

Pure, regular crystals can sometimes be obtained from Nature or man-made materials, such as samples of metals, minerals or other macroscopic materials.

The regularity of such crystals can sometimes be improved with annealing and other methods. However, in many cases, obtaining a diffraction-quality crystal is the chief barrier to solving its atomic-resolution structure.

Small-molecule and macromolecular crystallography differ in the range of possible techniques used to produce diffraction-quality crystals.

Small molecules generally have few degrees of conformational freedom, and may be crystallized by a wide range of methods, such as chemical vapour deposition and recrystallisation.

By contrast, macromolecules generally have many degrees of freedom and their crystallization must be carried out to maintain a stable structure.

For example, proteins and larger RNA molecules cannot be crystallized if their tertiary structure has been unfolded; therefore, the range of crystallization conditions is restricted to solution conditions in which such molecules remain folded.

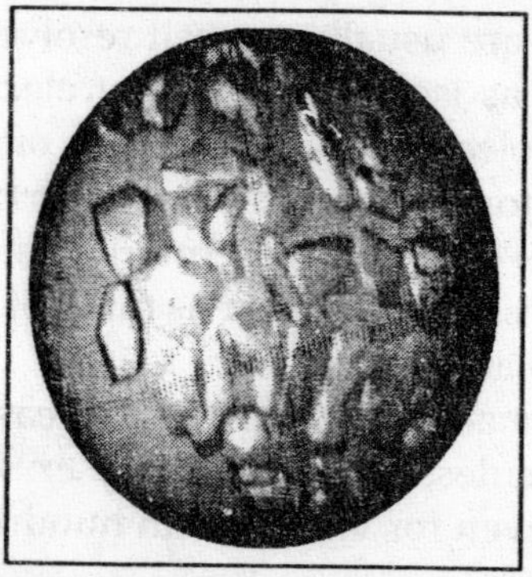

Fig. Protein Crystal seen Under a Microscope

Protein crystals are almost always grown in solution. The most common approach is to lower the solubility of its component molecules very gradually; however, if this is done too quickly, the molecules will precipitate from solution, forming a useless dust or amorphous gel on the bottom of the container. Crystal growth in solution is characterized by two steps: nucleation of a microscopic crystallite (possibly having only 100 molecules), followed by growth of that crystallite, ideally to a diffraction-quality crystal.

The solution conditions that favour the first step (nucleation) are not always the same conditions that favour the second step. The crystallographer's goal is to identify solution conditions that favour the development of a single, large crystal, since larger crystals offer improved resolution of the molecule.

Consequently, the solution conditions should disfavor the first step (nucleation) but favour the second (growth), so that only one large crystal forms per droplet. If nucleation is favored too much, a shower of small crystallites will form in the droplet, rather than one large crystal; if favored too little, no crystal will form whatsoever.

It is extremely difficult to predict good conditions for nucleation or growth of well-ordered crystals. In practice, favorable conditions are identified by screening; a very large batch of the molecules is prepared, and a wide variety of crystallization solutions are tested. Hundreds, even thousands, of solution conditions are generally tried before finding one that succeeds in crystallizing the molecules.

The various conditions can use one or more physical mechanisms to lower the solubility of the molecule; for example, some may change the pH, some contain salts of the Hofmeister series or chemicals that lower the dielectric constant of the solution, and still others contain large polymers such as polyethylene glycol that drive the molecule out of solution by entropic effects.

It is also common to try several temperatures for encouraging crystallization, or to gradually lower the temperature so that the solution becomes supersaturated. These methods require large amounts of the target molecule, as they use high concentration of the molecule(s) to be crystallized.

Due to the difficulty in obtaining such large quantities (milligrams) of crystallisation grade protein, dispensing robots have been developed that are capable of accurately dispensing crystallisation trial drops that are of the order on 100 nanoliters in volume. This means that roughly 10-fold less protein is used per-experiment when compared to crystallisation trials setup by hand (on the order on 1 microliters).

Several factors are known to inhibit or mar crystallization. The growing crystals are generally held at a constant temperature and protected from shocks or vibrations that might disturb their crystallization. Impurities in the molecules or in the crystallization solutions are often inimical to crystallization. Conformational flexibility in the molecule also tends to make crystallization less likely, due to entropy.

Ironically, molecules that tend to self-assemble into regular helices are often unwilling to assemble into crystals. Crystals can be marred by twinning, which can occur when a unit cell can pack equally favorably in multiple orientations; although recent advances in computational methods have begun to allow the structures of twinned crystals to be solved, it is still very difficult. Having failed to crystallize a target molecule, a crystallographer may try again with a slightly modified version of the molecule; even small changes in molecular properties can lead to large differences in crystallization behaviour.

Mounting the Crystal

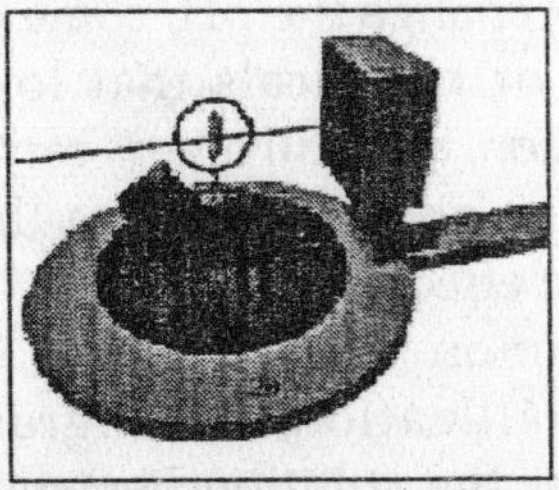

Once they are full-grown, the crystals are mounted so that they may be held in the X-ray beam and rotated. There are several methods of mounting. Although crystals were once loaded into glass capillaries with the crystallization solution (the mother liquor), a more modern approach is to scoop the crystal up in a tiny loop, made of nylon or plastic and attached to a solid rod, that is then flash-frozen with liquid nitrogen. This freezing reduces the radiation damage of the X-rays, as well as the noise in the Bragg peaks due to thermal motion (the Debye-Waller effect).

However, untreated crystals often crack if flash-frozen; therefore, they are generally pre-soaked in a cryoprotectant solution before freezing. Unfortunately, this pre-soak may itself cause the crystal to crack, ruining it for crystallography. Generally, successful cryo-conditions are identified by trial and error. The capillary or loop is mounted on a goniometer, which allows it to be positioned accurately within the X-ray beam and rotated. Since both the crystal and the beam are often very small, the crystal must be centered within the beam to within roughly 25 micrometres accuracy, which is aided by a camera focused on the crystal.

The most common type of goniometer is the "kappa goniometer", which offers three angles of rotation: the ω angle, which rotates about an axis roughly perpendicular to the beam; the κ angle, about an axis at roughly 50° to the ω axis; and, finally, the ϕ angle about the loop/capillary axis. When the κ angle is zero, the ù and axes are aligned. The κ rotation allows for convenient mounting of the crystal, since the arm in which the crystal is mounted may be swung out towards the crystallographer. The oscillations carried out during data collection involve the ù axis only. An older type of goniometer is the four-circle goniometer, and its relatives such as the six-circle goniometer.

X-ray Sources

The mounted crystal is then irradiated with a beam of monochromatic X-rays. The brightest and most useful X-ray sources are synchrotrons; their much higher luminosity allows for better resolution. They also make it convenient to tune the wavelength of the radiation, which is useful for multi-wavelength anomalous dispersion (MAD) phasing, described below. Synchrotrons are generally national facilities, each with several dedicated beamlines where data is collected around the clock, seven days a week.

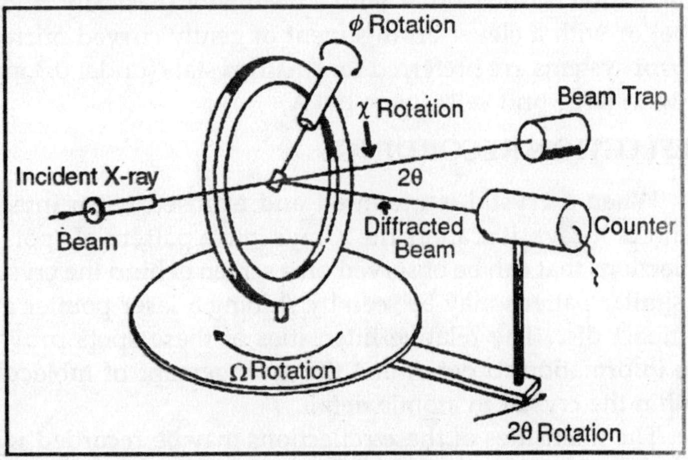

Fig. Diffractometer

Smaller, weaker X-ray sources are often used in laboratories to check the quality of crystals before bringing them to a synchrotron and sometimes to solve a crystal structure. In such systems, electrons are boiled off of a cathode and accelerated through a strong electric potential of roughly 50 kV; having reached a high speed, the electrons collide with a metal plate, emitting bremsstrahlung and some strong spectral lines corresponding to the excitation of inner-shell electrons of the metal.

The most common metal used is copper, which can be kept cool easily, due to its high thermal conductivity, and which produces strong Kα and Kβ lines. The Kβ line is sometimes suppressed with a thin layer (0.0005 in. thick) of nickel foil. The simplest and cheapest variety of sealed X-ray tube has a stationary anode (the Crookes tube) and produces circa 2 kW of X-ray radiation. The more expensive variety has a rotating-anode type source that produces circa 14 kW of X-ray radiation.

X-rays are generally filtered to a single wavelength (made monochromatic) and collimated to a single direction before they are allowed to strike the crystal. The filtering not only simplifies the data analysis, but also removes radiation that degrades the crystal without contributing useful information. Collimation is done either with a collimator (basically, a long tube) or with a clever arrangement of gently curved mirrors. Mirror systems are preferred for small crystals (under 0.3 mm) or with large unit cells (over 150 Å)

REFLECTION RECORDINGS

When a crystal is mounted and exposed to an intense beam of X-rays, it scatters the X-rays into a pattern of spots or reflections that can be observed on a screen behind the crystal. A similar pattern may be seen by shining a laser pointer at a compact disc. The relative intensities of these spots provide the information to determine the arrangement of molecules within the crystal in atomic detail.

The intensities of these reflections may be recorded with photographic film, an area detector or with a charge-coupled device (CCD) image sensor. The peaks at small angles correspond to low-resolution data, whereas those at high angles represent high-resolution data; thus, an upper limit on the eventual resolution of the structure can be determined from the first few images. Some measures of diffraction quality can be determined at this point, such as the mosaicity of the crystal and its overall disorder, as observed in the peak widths. Some pathologies of the crystal that would render it unfit for solving the structure can also be diagnosed quickly at this point.

One image of spots is insufficient to reconstruct the whole crystal; it represents only a small slice of the full Fourier transform. To collect all the necessary information, the crystal must be rotated step-by-step through 180°, with an image recorded at every step; actually, slightly more than 180° is required to cover reciprocal space, due to the curvature of the Ewald sphere.

However, if the crystal has a higher symmetry, a smaller angle such as 90° or 45° may be recorded. The axis of the rotation should generally be changed at least once, to avoid developing a "blind spot" in reciprocal space close to the rotation axis. It is customary to rock the crystal slightly (by 0.5-2°) to catch a broader region of reciprocal space.

Multiple data sets may be necessary for certain phasing methods. For example, MAD phasing requires that the scattering be recorded at at least three (and usually four, for redundancy) wavelengths of the incoming X-ray radiation. A single crystal may degrade too much during the collection of one data set, owing to radiation damage; in such cases, data sets on multiple crystals must be taken.

THEORY OF DIFFRACTION

The main goal of X-ray crystallography is to determine the density of electrons f(r) throughout the crystal, where r represents the three-dimensional position vector within the crystal. To do this, X-ray scattering is used to collect data about its Fourier transform F(q), which is inverted mathematically to obtain the density defined in real space, using the formula

$$f(r) = \int \frac{dq}{(2\pi)^3} F(q) e^{iq-r}$$

where the integral is summed over all possible values of q. The three-dimensional real vector q represents a point in reciprocal space, that is, to a particular oscillation in the electron density as one moves in the direction in which q points. The length of q corresponds to 2ð divided by the wavelength of the oscillation. The corresponding formula for a Fourier transform will be used below.

$$f(q) = \pm \int drf(r)e - iq \cdot r$$

where the integral is summed over all possible values of the position vector r within the crystal.

The Fourier transform F(q) is generally a complex number, and therefore has a magnitude |F(q)| and a phase ϕ(q) related by the equation,

$$F(q) = |F(q)|e^{if(q)}$$

The intensities of the reflections observed in X-ray diffraction give us the magnitudes |F(q)| but not the phases $\phi(q)$. To obtain the phases, full sets of reflections are collected with known alterations to the scattering, either by modulating the wavelength past a certain absorption edge or by adding strongly scattering (i.e., electron-dense) metal atoms such as mercury. Combining the magnitudes and phases yields the full Fourier transform F(q), which may be inverted to obtain the electron density $f(r)$.

Crystals are often idealized as being perfectly periodic. In that ideal case, the atoms are positioned on a perfect lattice, the electron density is perfectly periodic, and the Fourier transform $F(q)$ is zero except when q belongs to the reciprocal lattice (the so-called Bragg peaks). In reality, however, crystals are not perfectly periodic; atoms vibrate about their mean position, and there may be disorder of various types, such as mosaicity, dislocations, various point defects, and heterogeneity in the conformation of crystallized molecules. Therefore, the Bragg peaks have a finite width and there may be significant diffuse scattering, a continuum of scattered X-rays that fall between the Bragg peaks.

BRAGG'S LAW

An intuitive understanding of X-ray diffraction can be obtained from the Bragg model of diffraction. In this model, a given reflection is associated with a set of evenly spaced sheets running through the crystal, usually passing through the centers of the atoms of the crystal lattice. The orientation of a particular set of sheets is identified by its three Miller indices (h, k, l), and let their spacing be noted by d. William Lawrence Bragg proposed a model in which the incoming X-rays are scattered specularly (mirror-like) from each plane; from that assumption, X-rays scattered from adjacent planes will combine constructively (constructive interference) when the angle θ between the plane and the X-ray results in a path-length difference that is an integer multiple n of the X-ray wavelength λ.

$$2d \sin \theta = n\lambda$$

A reflection is said to be indexed when its Miller indices have been identified from the known wavelength and the scattering angle 2θ. Such indexing gives the unit-cell parameters, the lengths and angles of the unit-cell, as well as its space group. Since Bragg's law does not interpret the relative intensities of the reflections, however, it is generally inadequate to solve for the arrangement of atoms within the unit-cell; for that, a Fourier transform method must be carried out.

SCATTERING AS A FOURIER TRANSFORM

The incoming X-ray beam has a polarization and should be represented as a vector wave; however, for simplicity, let it be represented here as a scalar wave. We also ignore the complication of the time dependence of the wave and just focus on the wave's spatial dependence. Plane waves can be represented by a wave vector kin, and so the strength of the incoming wave at time t=0 is given by,

$$Ae^{ik_{in}\text{-}r}$$

At position r within the sample, let there be a density of scatterers $f(r)$; these scatterers should produce a scattered spherical wave of amplitude proportional to the local amplitude of the incoming wave times the number of scatterers in a small volume dV about r amplitude of scattered wave = $Ae^{ik\text{-}r}\ Sf(r)dV$ where S is the proportionality constant.

Let's consider the fraction of scattered waves that leave with an outgoing wave-vector of kout and strike the screen at rscreen. Since no energy is lost (elastic, not inelastic scattering), the wavelengths are the same as are the magnitudes of the wave-vectors |kin| = |kout|. From the time that the photon is scattered at r until it is absorbed at rscreen, the photon undergoes a change in phase

$$e^{ik}_{out}.(r_{screen} - r)$$

The net radiation arriving at rscreen is the sum of all the scattered waves throughout the crystal,

$$AS \int dr\ f(r) e^{ik_{in}.r} e^{ik_{out}.(r_{screen}-r)}$$

$$= ASe^{ik_{out}.r_{screen}} \int dr\ f(r) e^{i(k_{in}-k_{out}).r}$$

which may be written as a Fourier transform,

$$ASe^{ik_{out}.r_{screen}} \int dr\ f(r) e^{-iq.r}$$

$$= ASe^{ik_{out}.r_{screen}} F(q)$$

where $q = k_{out} - k_{in}$. The measured intensity of the reflection will be square of this amplitude

$$A^2S^2\ |F(q)|^2$$

Friedel and Bijvoet Mates

For every reflection corresponding to a point q in the reciprocal space, there is another reflection of the same intensity at the opposite point $-q$. This opposite reflection is known as the Friedel mate of the original reflection. This symmetry results from the mathematical fact that the density of electrons $f(r)$ at a position r is always a real number. As noted above, $F(r)$ is the inverse transform of its Fourier transform $F(q)$; however, such an inverse transform is a complex number in general. To ensure that $f(r)$ is real, the Fourier transform $F(q)$ must be such that the Friedel mates $F(-q)$ and $F(q)$ are complex conjugates of one another. Thus, $F(-q)$ has the same magnitude as $F(q)$— that is, $|F|(q) = |F|(-q)$—but they have the opposite phase, i.e., $\phi(q) = -\phi(q)$

$$F(-q) \quad = |F(-q)|\ e^{i\phi(-q)} = F^*(q) = |F(q)| e^{-i\phi(q)}$$

The equality of their magnitudes ensures that the Friedel mates have the same intensity |F|2. This symmetry allows one to measure the full Fourier transform from only half the reciprocal space, e.g., by rotating the crystal slightly more than a 180°, instead of a full turn. In crystals with significant symmetry, even more reflections may have the same intensity (Bijvoet mates); in such cases, even less of the reciprocal space may need to be measured, e.g., slightly more than 90°.

The Friedel-mate constraint can be derived from the definition of the inverse Fourier transform,

$$f(r) = \int \frac{dq}{(2\pi)^3} F(q) e^{iq.r}$$

$$= \int \frac{dq}{(2\pi)^3} |F(q)|\, e^{i\phi(q)} e^{iq.r}$$

Since Euler's formula states that eix = cos(x) + i sin(x), the inverse Fourier transform can be separated into a sum of a purely real part and a purely imaginary part,

$$f(r) = \int \frac{dq}{(2\pi)^3} |F(q)| e^{i(\phi + q.r)}$$

$$= \int \frac{dq}{(2\pi)^3} |F(q)| \cos(\phi + q.r)$$

$$+ i \int \frac{dq}{(2\pi)^3} |F(q)| \sin(\phi + q.r)$$

$$= I_{\cos} + iI_{\sin}$$

The function f(r) is real if and only if the second integral Isin is zero for all values of r. In turn, this is true if and only if the above constraint is satisfied

$$Isin = \int \frac{dq}{(2\pi)^3} |F(q)| \sin(\phi + q.r)$$

$$= \int \frac{dq}{(2\pi)^3} |F(-q)| \sin(-\phi - q.r) = -I_{\sin}$$

since $I_{\sin}$ = "$I_{\sin}$ implies that $I_{\sin} = 0$.

EWALD'S SPHERE

Each X-ray diffraction image represents only a slice, a spherical slice of reciprocal space, as may be seen by the Ewald sphere construction. Both kout and kin have the same length, due to the elastic scattering, since the wavelength has not changed. Therefore, they may be represented as two radial vectors in a sphere in reciprocal space, which shows the values of q that are sampled in a given diffraction image. Since there is a slight spread in the incoming wavelengths of the incoming X-ray beam, the values of $|F(q)|$ can be measured only for q vectors located between the two spheres corresponding to those radii.

Therefore, to obtain a full set of Fourier transform data, it is necessary to rotate the crystal through slightly more than 180°, or sometimes less if sufficient symmetry is present. A full 360° rotation is not needed because of a symmetry intrinsic to the Fourier transforms of real functions (such as the electron density), but "slightly more" than 180° is needed to cover all of reciprocal space within a given resolution because of the curvature of the Ewald sphere (add Figure to illustrate this). In practice, the crystal is rocked by a small amount (0.25-1°) to incorporate reflections near the boundaries of the spherical Ewald shells.

Patterson function

A well-known result of Fourier transforms is the autocorrelation theorem, which states that the autocorrelation $c(r)$ of a function $f(r)$

$$c(r) = \int dx f(x) f(x+r)$$

$$= \int \frac{dq}{(2\pi)^3} C(q) e^{iq.r}$$

has a Fourier transform $C(q)$ that is the squared magnitude of $F(q)$

$$C(q) = |F(q)|^2$$

Therefore, the autocorrelation function c(r) of the electron density (also known as the Patterson function) can be computed directly from the reflection intensities, without computing the phases. In principle, this could be used to determine the crystal structure directly; however, it is difficult to realise in practice. The autocorrelation function corresponds to the distribution of vectors between atoms in the crystal; thus, a crystal of N atoms in its unit cell may have N(N-1) peaks in its Patterson function. Given the inevitable errors in measuring the intensities, and the mathematical difficulties of reconstructing atomic positions from the interatomic vectors, this technique is rarely used to solve structures, except for the simplest crystals.

ADVANTAGES OF A CRYSTAL

In principle, an atomic structure could be determined from applying X-ray scattering to non-crystalline samples, even to a single molecule. However, crystals offer a much stronger signal due to their periodicity.

A crystalline sample is by definition periodic; a crystal is composed of many unit cells repeated indefinitely in three independent directions. Such periodic systems have a Fourier transform that is concentrated at periodically repeating points in reciprocal space known as Bragg peaks; the Bragg peaks correspond to the reflection spots observed in the diffraction image.

Since the amplitude at these reflections grows linearly with the number N of scatterers, the observed intensity of these spots should grow quadratically, like N^2. In other words, using a crystal concentrates the weak scattering of the individual unit cells into a much more powerful, coherent reflection that can be observed above the noise. This is an example of constructive interference.

In a non-crystalline sample, molecules within that sample would be in random orientations and therefore would have a continuous Fourier spectrum that spreads its amplitude more uniformly and with a much reduced intensity, as is observed in SAXS. More importantly, the orientational information is lost. In the crystal, the molecules adopt the same orientation within the crystal, whereas in a liquid, powder or amorphous state, the observed signal is averaged over the possible orientations of the molecules.

Although theoretically possible with sufficiently low-noise data, it is generally difficult to obtain atomic-resolution structures of complicated, asymmetric molecules from such rotationally averaged scattering data. An intermediate case is fibre diffraction in which the subunits are arranged periodically in at least one dimension.

X-RAY POWDER DIFFRACTION

ANALYTICAL METHODS

X-rays were discovered by W.C. Röentgen in 1895, and led to three major uses:

- X-ray radiography is used for creating images of light-opaque materials. It relies on the relationship between density of materials and absorption of x-rays. Applications include a variety of medical and industrial applications.
- X-ray crystallography relies on the dual wave/particle nature of x-rays to discover information about the structure of crystalline materials.
- X-ray fluorescence spectrometry relies on characteristic secondary radiation emitted by materials when excited by a high-energy x-ray source and is used primarily to determine amounts of particular elements in materials.

This course is primarily concerned with the x-ray crystallography of powders. In course materials you will commonly find X-ray Diffraction, X-ray powder diffraction, and the abbreviation XRD used interchangeably. Though intellectually somewhat sloppy, it is also common practice.

USES

The most widespread use of x-ray powder diffraction, and the one we focus on here, is for the identification of crystalline compounds by their diffraction pattern. Listed below are some specific uses that we will cover in this course:

- Identification of single-phase materials – minerals, chemical compounds, ceramics or other engineered materials.
- Identification of multiple phases in microcrystalline mixtures (i.e., rocks)

- Determination of the crystal structure of identified materials
- Identification and structural analysis of clay minerals
- Recognition of amorphous materials in partially crystalline mixtures

Below are some more advanced techniques. Some of these will be addressed in an introductory fashion in this course. Many are left for more advanced individual study.

- Crystallographic structural analysis and unit-cell calculations for crystalline materials.
- Quantitative determination of amounts of different phases in multi-phase mixtures by peak-ratio calculations.
- Quantitative determination of phases by whole-pattern refinement.
- Determination of crystallite size from analysis of peak broadening.
- Determine of crystallite shape from study of peak symmetry.
- Study of thermal expansion in crystal structures using in-situ heating stage equipment.

MONOCHROMATIC X RAY

The physics and mathematics describing the generation of monochromatic X-rays, and the diffraction of those X-rays by crystalline powders are very complex (and way beyond my limited abilities to expound upon them). Fortunately a complete understanding of the mathematics involved is not required to obtain, interpret and use XRD data. What is required is a basic understanding of how the X-rays interact with your specimen, the sources and characteristics of possible errors, and what the data tell you about your sample1.

What follows is a generalized explanation of the process of going from X-rays to diffraction data for math-challenged geologists like me. Some of these processes will be treated a bit more rigorously later in the course. For those who want to delve into the physics of X-ray diffraction. The intent here is to provide a conceptual framework for what is happening.

Below is a schematic diagram of a diffractometer system and on the next page is a photograph of our Scintag PAD V goniometer with many of the parts discussed below labeled.

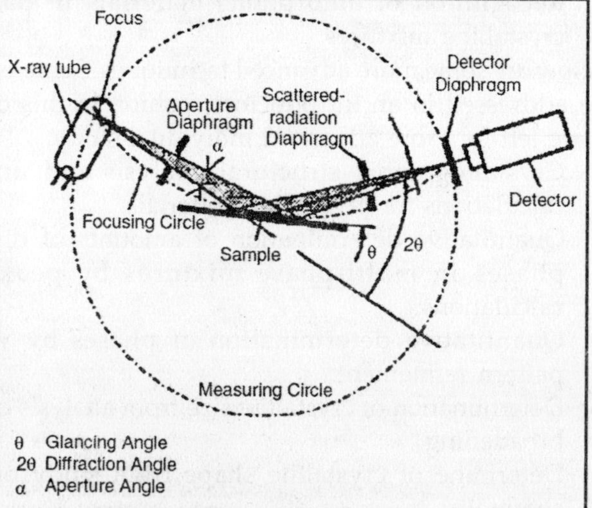

Diffractometer Beam Path in θ/2θ Mode

- The X-ray tube
- The flat specimen (labeled sample in the diagram)
- The goniometer circle (labeled measuring circle in the diagram) which remains constant through the analysis and is defined by the position of the target (Cu in our system) in the X-ray tube, the centre of the sample, and the position of the receiving slit (labeled detector diaphragm) on the detector side.
- The X-ray tube, specimen and receiving slit also lie on the arc of the focusing circle. Unlike the goniometer circle which remains fixed, the radius of the focusing circle is a function of θ - 2θ, with the radius decreasing as θ increases.
- The incident angle θ defined as the angle between the incident beam and the sample, and 2θ defined as the angle between the incident and diffracted beams. The detector is moved (rotated) at twice the angular rate of the sample to maintain the θ-2θ geometry.

- A filter (on the diffracted beam side) is used (in this example) to remove all but the desired Kα radiation from the diffracted beam before it enters the detector.
- A slit (labeled aperture diaphragm) on the incident beam side is used to narrow the beam so that it is confined within the area of the specimen.
- The path AB=BC is the radius of the diffractometer circle.
- The tube position is fixed and the θ-2θ geometry is maintained by rotating the sample holder at ½ the angular rate of the detector.
- There are Soller slits on both the tube and detector side, and two collimating and receiving slits.
- Note the easy-to-read angular indicators and micrometer dials for visually reading θ and 2θ.
- The detector on this system also includes a graphite monochromator adjacent to the scintillation detector (off the photo, top right) eliminating the need for any filters in the system.

Preparation

The Ideal Specimen is a statistically infinite amount of randomly oriented powder with crystallite size less than 10 μm, mounted in a manner in which there is no preferred crystallite orientation. In this day of automated data collection and analysis, the preparation of your specimen is usually the most critical factor influencing the quality of your analytical data. Sample preparation is a significant topic in this course.

GENERATE ANALYTICAL X-RAYS

A coherent beam of monochromatic X-rays of known wavelength is required for XRD analysis Striking a pure anode of a particular metal with high-energy electrons in a sealed vacuum tube generates X-rays that may be used for X-ray diffraction. By the right choice of metal anode and energy of accelerated electrons, a known wavelength (i.e., energy) or group of wavelengths will dominate the X-rays generated. Copper (Cu) X-ray tubes are most commonly used for X-ray diffraction of inorganic materials. The wavelength of the strongest Cu radiation (Kα) is approximately 1.54 angstroms (Å). Other anodes commonly used in X-ray generating tubes include Cr (Kα 2.29 Å), Fe (Kα 1.94 Å), Co (Kα 1.79 Å), and Mo (Ká 0.71 Å).

The full spectrum of radiation produced, and how it is "processed" to get to a (more or less) monochromatic character will be discussed in more detail later. For most X-ray diffraction applications, the closer we can get to monochromatic radiation in our X-ray beam, the better our experimental results will be.

The radiation produced in the tube includes $K\alpha_1$, $K\alpha_2$, and Kβ as the highest energy X-rays and a whole host of lower energy radiation. We generally use the Kα for our analytical work. The Kβ radiation is usually removed by use of a filter, a monochromator or an energy-selective detector. The $K\alpha_2$ radiation is removed from the X-ray data electronically during data processing.

DIRECT THE X-RAYS AT A POWDERED SPECIMEN

An approximately parallel beam of X-rays is directed at the powdered specimen. In most powder diffractometers systems a series of parallel plates (soller slits) arranged parallel to the plane of the diffractometer circle and several scatter and receiving slits (arranged perpendicular to the diffractometer circle) are used to create an incident beam of X-rays that are (approximately) parallel.

Soller slits are commonly used on both the incident and diffracted beam, but this will vary depending on the particular system.

The scatter slits (on the incident beam side) may be varied to control the width of the incident beam that impinges upon the specimen and the receiving slits may be varied to control the width of the beam entering the detector. Filters for removing Kâ may be located in the beam path on the generator or detector side of the path; a monochromator, if present, is usually located on the detector side between the receiving slit and the detector.

The newest diffractometers may be purchased with sophisticated (usually of a very proprietary nature) optical systems that can create a tightly controlled and focused incident beam of X-rays.

MEASURING X-RAYS

Interaction of X-rays with sample creates secondary "diffracted" beams (actually generated in the form of cones) of X-rays related to interplanar spacings in the crystalline powder according to a mathematical relation called "Bragg's Law":

$$n\lambda = 2d \sin\theta$$

Where:

n is an integer

λ is the wavelength of the X-rays

d is the interplanar spacing generating the diffraction and

θ is the diffraction angle

λ and d are measured in the same units, usually angstroms. We will derive the Bragg law a bit more rigorously later but for a powder specimen in a diffractometer having a statistically infinite amount of randomly oriented crystallites, diffraction maxima (or peaks) are measured along the 2θ diffractometer circle.

Powder diffractometers come in two basic varieties: θ-θin which the X-ray tube and detector move simultaneously or a θ-2θ in which the X-ray tube is fixed, and the specimen moves at ½ the rate of the detector to maintain the θ--2θ geometry. Our Scintag PAD V system is a θ-2θ system; the Siemens D5000 systems located in the Chemistry Department are θ-θ systems. In both systems the geometry shown in the previous diagrams is maintained during data collection.

The "angle" of the diffraction (recorded as 2θ by convention) is related to the interplanar spacing, d, by the Bragg law, and the intensity of the diffraction maximum is related to the strength of those diffractions in the specimen. The angles and intensities of diffractions are recorded electronically using a detector, electronics and specialized software resulting in a plot of 2θ (horizontal axis) vs. intensity (vertical axis) for the specimen.

DETECTORS

There are a variety of detectors used in XRD systems. The Scintag system in our laboratory uses a scintillation counter. In the Chemistry XRD lab, the Siemens systems have either scintillation counters or a large-window position sensitive detector or PSD (covering 8º 2θ simultaneously). Scintillation counters are some of the oldest technology available, but are still widely used because of their relatively low cost, ease of use and durability. Newer detector technologies can deliver improved quality data, a higher degree of energy selectivity or deliver it faster. All detector technologies also involve tradeoffs in cost and/or maintenance. We will briefly discuss detector technologies later in the course.

"Legacy" Methods

Before the advent of computerized data collection, X-ray diffraction data were derived by film methods or by diffractometers using paper strip-chart recorders. In both cases, the resultant data were derived by physically measuring peak positions and intensities, and the diffraction data recorded as a list of peaks (in degrees 2θ) and relative intensities (scaled from 0 to 100). Modern automated diffractometers and the associated automation software collect data electronically, process and calculate it digitally removing much of the tedium from the acquisition of powder XRD data. Some of the "legacy" methods are useful for understanding the process of diffraction, and we will address some of these in this course.

DETERMINATION OF CRYSTALLINE PHASES

For most samples, the aim of the analysis is to identify the crystalline phases present. Even for work where other information is sought (i.e., unit cell calculations, quantitative determinations, etc.), identification of the phase(s) present is usually the first step. Phase identification is accomplished by comparing the data (peaks and relative intensities) from your specimen with peaks and relative intensities from a very large set of "standard" data provided by the International Centre for Diffraction Data (ICDD). The current PDF4 release contains

254,873 Digital XRD patterns, both experimental and calculated, from almost every known inorganic and many organic crystalline substances. In our lab we use Jade to facilitate the access to this massive (and continually growing) database. Jade includes an automated search-match function that compares the sample pattern with the ICDD database. With good data from a single-phase sample, Jade's automated search-match programme will usually identify the phase successfully with little or no effort on your part. For most two-phase samples identification of the dominant phase will usually be successful, but the second may require more hunting.

With three or more phases (and virtually all bulk rocks), some knowledge of the likely constituents will be required to successfully "sleuth" the constituents. Fortunately the ability to visually compare your sample pattern to a large number of possible phases is a manageable task. We will spend considerable laboratory time learning to use this powerful software. Prior to the advent of automated XRD software like Jade, manual methods required the listing of all the 2θ-intensity values for your sample and the use of paper indexes to identify phases.

Chapter 3

Chemical Bonding

DALTON ATOMIC THEORY

Before Dalton's atomic theory there wasn't much reason to consider how atoms stick together, although it was clear that some force is necessary to bind compounds of elements together. The term "chemical affinity" referred more to the ability to react (undergo a chemical change) than to form a stable compound. When it was realized that chemical compounds were formed of discrete molecules containing fixed numbers of atoms, the need to explain molecular stability developed, leading to the notion of the interatomic chemical bond (Latin inter for among plus band for a fastening) holding atoms together. Early Nineteenth Century speculations about the origin of interatomic bonds were affected by contemporary discoveries in electricity, indicating that "opposites" attract.

Positive and negative charges are electrical opposites, and north and south poles are magnetic opposites. Precisely just what constituted chemical opposites wasn't clear, however. Chemists had classified substances into metals and non-metals, and into acids and bases, both of which seemed opposite in the sense that they had "affinity" for each other and readily reacted together to form new compounds. Yet there was little evidence to support the hypothesis of Avogadro that non-metal gases were composed of two atoms of the same element, nor were there any explanations for the solid state of metallic elements, or the extreme hardness of non-metallic elements such as diamond.

The number of bonds an element could make, called the

valence (Latin valere for strength) of the element, is suggested by the atom ratios of chemical formulas, and ordinarily limited to a small number. For example, hydrogen atoms never forms bonds to more than one another atom, so are assigned a valency of one. Oxygen has a valency of two (is divalent) as each oxygen atom combines with two hydrogen atoms in forming water (H_2O, more descriptively written showing atom connections as H-O-H). Carbon has a valency of four (is tetravalent) because it combines with a maximum of four hydrogen atoms (in methane, CH4 more descriptively shown with C surrounded by 4 H), or, equivalently, a maximum of two oxygen atoms, each with a valency of two (in carbon dioxide, CO_2, more descriptively written O-C-O).

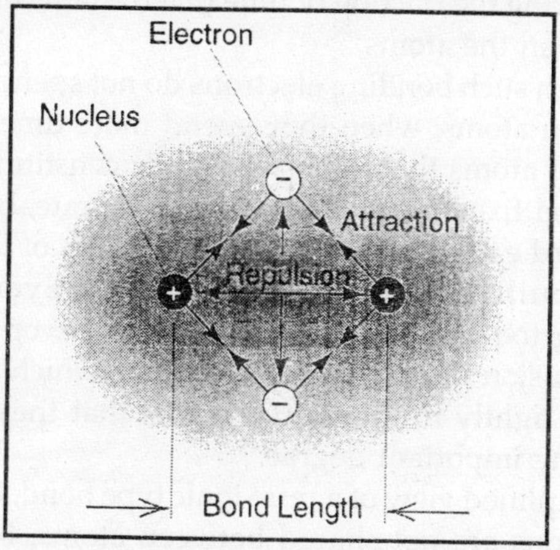

Fig. Chemical Bonds between Atoms

The electrons of atoms are electromagnetically attracted by the nuclei of atoms, due to the opposite electric charge of electrons and nuclei. Chemical bonds are characterised by physical states in which a few electrons move partly from one atom to one or more other atoms, driven by the achievement of a lower state of energy from this motion. This lowering of energy is caused by a rearrangement of charges, usually resulting in net decrease in the average distance between the electrons of all the bonded atoms, and their nuclei. The transfer of charge caused by the movement of the electron from one atom to another, also causes the participating atoms (which may number from two to many) to be attracted to one another electromagnetically. The attractive force between atoms is the bond. Chemical bonds, for the sake of simplicity, are classically assigned characteristics of two major types: covalent and ionic.

In a simplified view of a pure covalent type bond, the bond forms as a few electrons farthest from their atomic nuclei become more attracted to the region of space between two nuclei. In this region, negatively-charged electrons experience attraction from the positively-charged protons from more than one nucleus. This causes some electrons to spend a high probability of their time in the interatomic space. In turn, the nuclei are stabilized in position by the pull from these shared electrons during the fraction of time that the bonding electrons reside between the atoms.

Although such bonding electrons do not spend all of their time between atoms, when they spend more time between a given pair of atoms than otherwise, they constitute chemical bonds. Nuclei fixed by such bonds may vibrate, but they are pulled toward each other by the mutual forces of the bonding electrons pulling them together, yet prevented from approaching too closely by their own charge, or else by the mutual repulsion of other inner electrons, which are held so closely and tightly to individual nuclei that they cannot be shared to any important degree.

In a simplified view of a pure ionic type bond, one or more outer electrons are not shared between atoms, but donated from one atom to another. In such a bond, the structure of the electron cloud of one of the nuclei contains an available space for another electron, which allows an additional electron to experience a greater net attraction from the nucleus than is experienced by outer electrons in a neighboring atom toward their own nucleus. This difference in available states causes effective transfer of one or more electrons from one atom to another atom, where they can be more tightly bound. This transfer causes the donating atom to assume a net positive charge, and the other to assume a net negative charge; the atoms thus become positive or negatively charged ions. The bond then results from electrostatic attraction between these ionized atoms. Most bonds have a mixture of covalent and ionic character, as bonding electrons are shared between atoms, but shared somewhat unevenly. All bonds can be explained by quantum theory, but in practice, simplification rules allow chemists to predict the strength, directionality, and polarity of bonds. The octet rule and VSEPR theory are two examples. More sophisticated theories are valence bond theory which includes orbital hybridization and resonance, and the linear combination of atomic orbitals molecular orbital method which includes ligand field theory. Electrostatics are used to describe bond polarities and the effects they have on chemical substances.

IONIC AND COVALENT BONDS

Chemical salts are found to conduct electricity when melted or dissolved in water, suggesting the presence of charge carriers, called ions (from the Greek for traveller). Since salts are neutral compounds, the charge carriers must be divided into two groups, one positively charged and the other negatively charged. The oppositely charged ions would be bound through electrostatic attractions, or ionic bonds. Because chemical salts are known to be composed of metallic elements bound to non-metallic elements, the most reactive of which are found at opposite ends of the periodic table, ionic attraction can be explained in terms of the attraction of opposites.

When metals or nonmetals bond to elements of their own kind, the reason for the bonding is not so obvious. Compounds containing different elements might have some attraction due to the differences between the elements, but homonuclear (one-element) polyatomic (more than one atom) elements such as H2 or sodium metal are bound by some common feature rather than by some difference between the atoms. Such bonds, called covalent bonds (Latin co + valere for joint strength), are not limited to homonuclear molecules but also occur between atoms of different elements that are close in behaviour. The periodic table summarizes the properties of the elements, including bonding trends as well, and can be used to decide the basic type of bonding between atoms in molecules in a binary.

IONIC BONDS

Ionic bonds are simply the attractive forces between oppositely-charged ions. Ionic compounds contain cations (which can be either monatomic or polyatomic) and anions (which can be either monatomic or polyatomic.) Remember that ions are formed by the loss or gain of electrons. For monatomic cations, the electrons which are lost first in the formation of the ion are always the outermost (valence) electrons. For example, you already know that the 4s subshell is at lower energy than the 3d subshell in a neutral atom. This is why the 4s subshell fills before the 3d. For ions, however, the order of orbital energies approaches that for a hydrogen ion, where all subshells in a given main shell have the same energy.

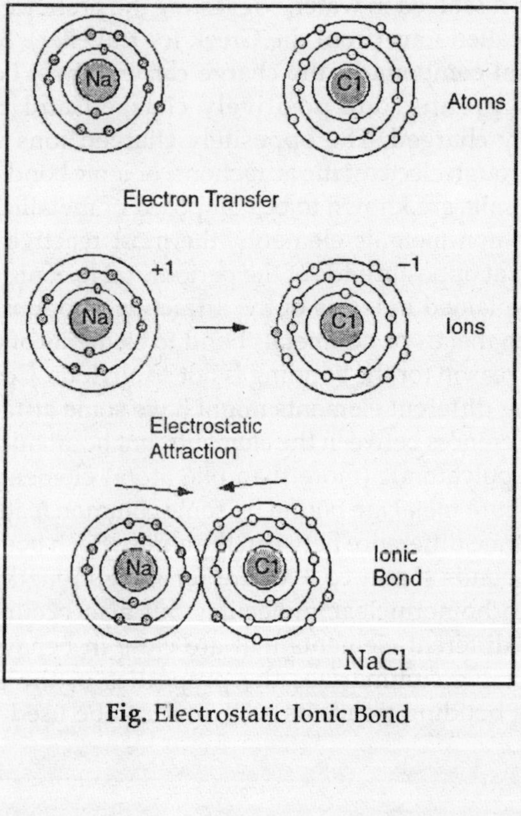

Fig. Electrostatic Ionic Bond

For this reason, the 4s electrons are lost before the 3d electrons when a cation is formed from a transition metal. For example, the configuration of a neutral cobalt atom is $[Ar]4s^2 3d^7$, but the configuration of a cobalt(III) ion is $[Ar]3d^6$. A simple view of the formation of an ionic compound is that the electrons lost by the cation are gained by the anion; in other words, a formal transfer of electrons can be envisioned. This is why the total positive charge from the cations must be balanced by the total negative charge from the anions - no free electrons are floating around. This gives rise to the formula Al_2O_3 for aluminum oxide rather than AlO_3, AlO, Al_3O_2, or other incorrect formulas.

ELECTRONEGATIVITY

Electronegativity is a measure of the tendency of an atom to attract the bonding electrons in a chemical bond toward itself. It is not a precisely-defined quantity like ionization potential or electron affinity, which can be measured accurately. In fact, there are more than a dozen scales of electronegativity. Luckily, for introductory courses only one scale is really needed to put the concept to use: this is the Pauling scale, named after the two-time Nobel prize winning chemist Linus Pauling. According to this scale, fluorine is the most electronegative element, with a Pauling electronegativity of 4.0. The least electronegative element is cesium, with a Pauling electronegativity of 0.7. Thus, there is a range of about 3.3 Pauling units between the electronegativities of the most and least electronegative elements. Be sure you understand the two periodic trends in electronegativity and the rule for getting the electronegativity of any second-period element just by memorizing the value of 4.0 for fluorine:

- Electronegativity increases from left to right in a given period.
- Electronegativity decreases from top to bottom in a given group.

To get the electronegativity of a second-period element, start at fluorine (4.0) and subtract 0.5 Pauling unit for each step to the left in the second period. Thus, oxygen has an electronegativity of 3.5, nitrogen has an electronegativity of 3.0, etc.

Sometimes a table of electronegativities is used to classify the possible bonds between each pair of atoms in a molecule into two categories, ionic or covalent. Bonds between atoms which differ in electronegativity on a scale of 0 to 4 by more than 2 units may be assumed to be ionic. However, since electronegativities of metals vary between 1 and 2 and non-metals between 2 and 4, metal to non-metal bonds are the only ionic bonds.

Electronegativities are useful for more refined classifications and trends in terms of the per cent ionic or covalent character of a bond. In these terms, ionic bonds are those with more than 50% ionic character.2 Because electronegativity is only a semiquantitative concept, there is no simple equation relating electronegativity to per cent ionic or covalent character, and there is no standard table of electronegativities.

This is due to the fact that atoms in molecules are only approximately localized and independent of the environment of the entire molecule. This prevents assigning unique electronegativities to atoms which experience a variety of bonding environments. A parameter used to indicate the charge an atom has or would have if it were ionized in compounds is the oxidation number. Three rules help define the assignment of oxidation numbers to atoms in compounds.

COVALENT BONDS

Covalent bonds are bonds formed by the sharing of electrons, in contrast to the transfer which can be envisioned for an ionic bond. Covalent bonds are formed when the electronegativity difference between the two elements involved in the bond is small, indicating that the two atoms have similar attractions for the electrons.

On the other hand, large differences in electronegativities give rise to ionic bonds. Half of the range (3.3) of Pauling electronegativity values is about 1.7. This value can be used as a ballpark dividing line between ionic and covalent bonds:

if the difference in electronegativities between the two atoms is greater than 1.7, then the bond is predominately ionic; if the difference is less than 1.7, then it is predominately covalent; if it is 1.7, then the bond has approximately 50% ionic and 50% covalent character.

It is unusual, however, that you'll need to make such quantitative determinations, especially since they are only approximate. For example, the electronegativity difference between Li and I is 2.5-1.0, or 1.5. This would imply more covalent than ionic character, but most chemists would classify lithium iodide as an ionic compound because it is composed of a metal and a nonmetal. Ionic compounds are formed between metals and nonmetals, whereas covalent compounds are formed between nonmetals. The simple reason is that metals are on the left side of the periodic table and nonmetals are on the right side, giving a fairly large difference in electronegativities. Nonmetals, on the other hand, are in the same general region of the periodic table and thus there are small electronegativity differences between two nonmetals.

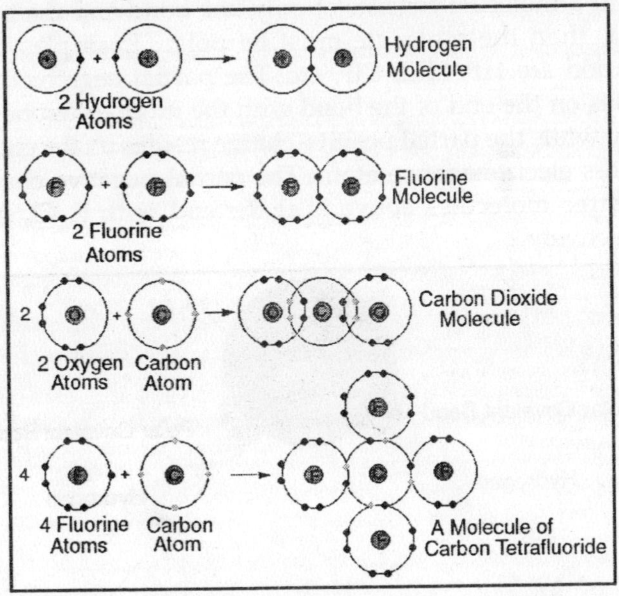

Fig. Covalent Bonds

The only purely covalent compounds are those where the electronegativity difference is zero - and this occurs for homonuclear diatomic molecules like N2, O2, H2, etc. One final clarification is necessary: the bonding within a polyatomic ion is covalent, since the atoms in the ion are nonmetals (NH4+, SO42-, etc.).

But compounds of these ions are ionic, since by definition an ionic bond is the electrostatic attraction between oppositely-charged ions in the crystal. Thus, sodium nitrate, NaNO3, is an ionic compound, but the bonding within the nitrate ion is covalent.

POLAR MOLECULES

You already know that covalent bonds represent shared electrons. This does not necessarily mean, however, that the electrons are shared equally. An unequal sharing of electrons results in a polar covalent bond (sometimes simply called a polar bond.) Such a bond results in a partial negative charge at one end of the bond and a partial positive charge at the other end. If a molecule consists of only one bond and the bond is polar, then the molecule must be polar. Examples of this situation are HF, HCl, ClF, etc. The partial negative charge resides on the end of the bond with the more electronegative atom, while the partial positive charge resides on the end with the less electronegative atom. The partial negative charge in the three molecules above is at the end with F, Cl, and F, respectively.

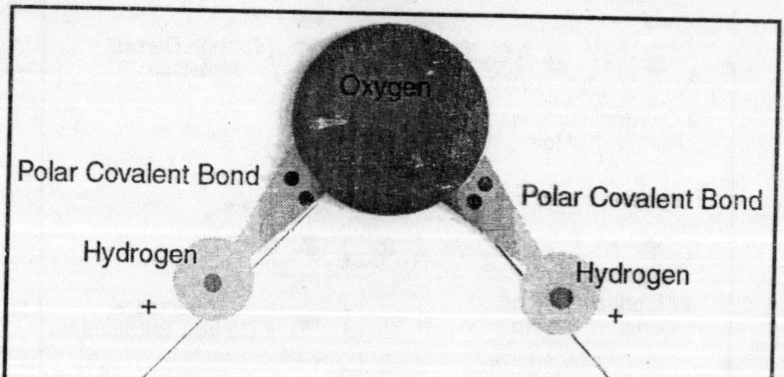

Fig. Polar Molecules of Covalent Bonds

If a molecule has more than one bond, then the molecule as a whole is nonpolar only if the individual polarities of the bonds cancel each other.

To decide if this happens, the geometry of the molecule must be known. For example, the carbon dioxide molecule is linear, with each C-O bond along the same line: O=C=O. Although each carbon-oxygen bond is polar, the molecule as a whole is nonpolar because the two equal polarities are in exactly opposite directions. Water, on the other hand, is a bent molecule.

The two H-O polar bonds do not cancel each other, and as a result the molecule is polar. Three molecules which have polar bonds but which have no molecular dipole moment because the individual bond polarities cancel one another.

COVALENT BOND ENERGIES

Bonds are stable because the ions in an ionic bond are at lower energy when they are close to one another in the crystal, and because the shared electrons in a covalent bond are attracted by the two nuclei of the bonded atoms. Thus, it takes energy to break ionic and covalent bonds.

Covalent bond energies are average energies, each of which depends on the chemical environment of the bonded atom.

For example, the successive bond dissociation energies for methane are shown on page 366 of the text. Although the four values (one for each C-H bond) are slightly different, their average value of 413 kJ/mol can be assumed to represent the energy of a covalent carbon-hydrogen bond.

A number of bond energies are shown in Table 8.4. Notice that the table is separated into single bond and multiple bond categories. Table 8.5 lists bond lengths for selected bonds. Notice that for a given pair of atoms, bond length decreases in the order,

single > double > triple.

Since the formation of covalent bonds releases energy and the breaking of bonds requires energy, we can calculate the approximate the enthalpy change for a chemical reaction by using tabulated bond energies. The procedure is simple: first, break apart all the bonds in the reactants and determine how much energy is required (this is endothermic, and thus a positive value.) Then determine how much energy is released when the bonds in the products form (this is exothermic, and thus a negative value.) Add the two quantities to get the enthalpy change for the reaction.

Important note: It is essential that the type of bonds in the reactants and products are known. For example, simply looking at the formula N2 might give one the impression that a single-bond N-N energy of 160 kJ per mole should be used. However, this will result in an incorrect answer, since there is a triple bond in the nitrogen molecule with a covalent bond energy of 941 kJ/mole.

The types of bonds in a molecule can be determined by drawing the Lewis electron-dot structure for the molecule.

Resonance

Although Lewis structures are very useful for a large number of molecules, there are some problems with using electron-dot structures to describe bonding:

One of the goals in drawing a Lewis structure is to pair up electrons. This is impossible for a molecule with an odd number of electrons, like NO or ClO_2. Which atom "has" the odd electron in these molecules?

Another goal for obtaining a valid Lewis structure is to obtain a duet of electrons around hydrogen or a completed octet around other atoms. The fact, however, is that there are exceptions to the octet rule: Be in BeF_2 and B in BF_3 have only four and six electrons in the Lewis structure, respectively, instead of eight. Atoms in the third period and below are able to "expand" their octet and can thus have more than eight electrons in the Lewis structure. Some examples are SF_4 (10 e- around sulfur) and XeF_4 (12 e- around xenon). There are many more examples.

Sometimes a single Lewis structure for a molecule doesn't accurately explain the properties of the molecule. For example, the Lewis structure for sulfur dioxide, SO_2, has a single sulfur-oxygen bond, a double sulfur-oxygen bond, and a lone pair of electrons on the central sulfur atom. All electrons are paired, 18 valence electrons are used in the Lewis structure, and each atom has an octet. Thus, the Lewis structure is "correct", at least in a formal sense. The problem is that the electron-dot diagram implies that there should be two bond strengths in sulfur dioxide - an S-O single bond and a stronger S-O double bond. Furthermore, there should also be two different bond lengths - a longer S-O single bond and a shorter S-O double bond. Here's the problem: experimentally, it has been found that both sulfur-oxygen bonds have the same bond length and the same bond strength. The concept of resonance is used to overcome this sort of problem with Lewis structures.

The problem mentioned above is not a problem with the molecule - it is a problem with the limited ability of a simple notation like a Lewis structure to describe a complicated concept like the bonding in a molecule. It is rather like trying to describe a real, three-dimensional object by drawing it on a two-dimensional piece of paper; the problem isn't with the object - it is with our limited resources for making the drawing.

The best way to describe the bonding in molecules like SO2 is to draw two or more Lewis structures and say that the actual molecule is a hybrid, or combination, of the separate structure; this is indicated by drawing double-headed arrows between the structures.

Each of the separate structures is called a resonance structure, and the actual molecule is described as a resonance hybrid of the resonance structures. It is important to remember two rules for drawing resonance structures:

The positions of all atoms in all resonance structures of a given molecule must be the same. In other words, atoms are not moved when a different resonance structure is drawn from an existing one; rather, electron pairs are moved.

Do not think that the actual structure results from an oscillation among the resonance structures. For example, each of the two resonance structures for sulfur dioxide has a single bond and a double bond: the difference is in the location of each. Don't think that at a particular instant there is a single bond between the sulfur and one of the oxygens and that in the next instant the bond becomes a double bond, with continual "flipping" back and forth. This is not the correct interpretation of resonance. Both bonds in SO_2 are identical, and the bonds are neither single nor double bonds. Each bond has properties intermediate between a single and double bond.

The resonance hybrid can be compared to a mule, which is a hybrid between a horse and a donkey. The mule has its own unique properties, and it exists in its own right — it isn't a horse half the time and a donkey the other half of the time.

Formal Charge

Unlike oxidation number, which can be determined from a formula like Na_2SO_4, the formal charge of an atom in a covalent molecule or polyatomic ion can only be determined from the Lewis structure. Formal charge is calculated by splitting the bonding electrons equally between the bonded atoms, and assigning all nonbonding electrons to the atom which "owns" them, then comparing the resulting number of electrons with the number of valence electrons the atom would have if it were neutral. Formal charges are shown as a positive number or negative number (e.g. +2 or -1) inside a circle and placed next to the atomic symbol in the Lewis structure. Recall that oxidation numbers are written without the circle.

For example, the Lewis structure for carbon dioxide shows two carbon-oxygen double bonds and two nonbonding pairs of electrons on each of the oxygen atoms. The carbon atom has four bonding pairs of electrons (two double bonds, or eight electrons) around it. The central carbon atom is assigned four of these eight electrons and each oxygen is assigned two of them. The formal charge of carbon is therefore number of valence electrons minus number of electrons after assignment, or 4 minus 4 = a formal charge of zero.

Each oxygen atom "owns" two nonbonding pairs of electrons (four electrons) and was assigned two of the bonding electrons, giving a total of six electrons. The formal charge is therefore 6-6= a formal charge of zero.

Now draw one of the resonance structures for sulfur

dioxide, with one S-O single bond, one S-O double bond, and a nonbonding pair on the central sulfur atom (remember to complete the octets around the oxygen atoms.) Assigning the electrons as described above leads to a formal charge of +1 for the sulfur, -1 for the singly-bonded oxygen and zero for the doubly-bonded oxygen.

Let's look at one more example. Draw the Lewis structure for the sulfate ion, $(SO_4)^{2-}$, which has four single sulfur-oxygen bonds. Verify the formal charges of +2 for the central sulfur atom and -1 for each of the oxygen atoms. Notice that in the CO_2, SO_2, and $(SO_4)^{2-}$ examples above, the algebraic sum of the formal charges is equal to the charge on the molecule or ion. This must always be the case. Remember that formal charges cannot be determined from a simple formula; the Lewis structure is needed.

Formal Charges and Resonance

Formal charges can be used to decide if a particular resonance structure is a "good" one - in other words, if the resonance structure contributes significantly to the actual structure.

Here are the rules:

- "Like" formal charges (plus/plus or minus/minus) should not reside on adjacent atoms.
- "Unlike" formal charges (plus/minus) should not be unnecessarily separated. Formal charges closer to zero usually represent more important resonance structures than those which have large formal charges.

For example, draw one of the Lewis structures for nitrous oxide (laughing gas, N_2O - the skeletal structure is N-N-O). You should now be able to "push" electron pairs around to come up with a total of three resonance structures which obey the octet rule.

Each of these resonance structures. Make sure you correctly determined these formal charges for all three of the resonance structures (Formal charges are listed in the following order: terminal nitrogen, central nitrogen, oxygen.)

N-N double bond, N-O double bond: formal charges are −1, +1, 0.

N-N triple bond, N-O single bond: formal charges are 0, +1, and -1.

N-N single bond, N-O triple bond: formal charges are −2, +1, and +1.

According to the rules mentioned above, resonance structures 1 and 2 are "good" structures and contribute significantly to the actual structure. Resonance structure 3 results in the placement of two positive formal charges on adjacent atoms, and thus does not contribute significantly to the actual structure.

VSEPR MODEL

According to the Valence Shell Electron Pair Repulsion model, the geometry of a molecule or polyatomic ion is determined by repulsions between valence electron pairs in the molecule. The idea is simple: since electrons all have the same charge, electron pairs repel each other, resulting in a geometry in which these mutual repulsions are minimized.

The overall geometry is determined by the total number of "groups" around the central atom in the species. A "group" is any of the following: a single bond, double bond, triple bond, or nonbonding pair.

The molecular geometry is obtained by first determining the overall geometry, then ignoring any nonbonding pairs around the central atom and deciding what shape results. A common abbreviation for representing a molecule or ion uses the symbol A for the central atom, X for an atom bonded to the central atom, and E for a nonbonding pair on the central atom.

Nonbonding pairs occupy equatorial (radial) positions in a trigonal bipyramid. Two nonbonding pairs in an overall octahedral structure occupy trans positions (across from one another). Bond angles decrease with increasing electronegativity of the non-central atoms, provided that there is at least one nonbonding pair in the species. For example, the F-N-F bond angle in NF3 is smaller than the H-N-H bond angle in ammonia. A simple way to understand this is to picture the bonding electrons in NF_3 as being closer to the more electronegative fluorine than the nitrogen, thereby allowing the nonbonding pair to "squeeze" the bonding pairs closer together, resulting in a smaller bond angle. In ammonia, the bonding electrons are closer to the more electronegative nitrogen than the hydrogen, and are therefore already close to one another. This doesn't allow the nonbonding pair to push the hydrogens as close together as it could for the fluorine atoms in NF3.

VALENCE BOND THEORY

Valence bond theory was formulated which argued essentially that a chemical bond forms when two valence electrons, in their respective atomic orbitals, work or function to hold two nuclei together, by virtue of system energy lowering effects. In 1931, building on this theory, chemist Linus Pauling published what some consider one of the most important papers in the history of chemistry: "On the Nature of the Chemical Bond". In this paper, building on the works of Lewis, and the valence bond theory (VB) of Heitler and London, and his own earlier work, he presented six rules for the shared electron bond, the first three of which were already generally known:

The electron-pair bond forms through the interaction of an unpaired electron on each of two atoms. The spins of the electrons have to be opposed. Once paired, the two electrons cannot take part in additional bonds. *His last three rules were new*: The electron-exchange terms for the bond involves only one wave function from each atom. The available electrons in the lowest energy level form the strongest bonds.

Of two orbitals in an atom, the one that can overlap the most with an orbital from another atom will form the strongest bond, and this bond will tend to lie in the direction of the concentrated orbital. Building on this article, Pauling's 1939 textbook: On the Nature of the Chemical Bond would become what some have called the "bible" of modern chemistry. This book helped experimental chemists to understand the impact of quantum theory on chemistry. However, the later edition in 1959 failed to address adequately the problems that appeared to be better understood by molecular orbital theory. The impact of valence theory declined during the 1960s and 1970's as molecular orbital theory grew in popularity and was implemented in many large computer programs. Since the 1980s, the more difficult problems of implementing valence bond theory into computer programs have been largely solved and valence bond theory has seen a resurgence.

MOLECULAR ORBITAL THEORY

Molecular orbital theory (MO) uses a linear combination of atomic orbitals to form molecular orbitals which cover the whole molecule. These are often divided into bonding orbitals, anti-bonding orbitals, and non-bonding orbitals. A molecular orbital is merely a Schrödinger orbital which includes several, but often only two nuclei. If this orbital is of type in which the electron(s) in the orbital have a higher probability of being between nuclei than elsewhere, the orbital will be a bonding orbital, and will tend to hold the nuclei together.

If the electrons tend to be present in a molecular orbital in which they spend more time elsewhere than between the nuclei, the orbital will function as an anti-bonding orbital and will actually weaken the bond.

Electrons in non-bonding orbitals tend to be in deep orbitals (nearly atomic orbitals) associated almost entirely with one nucleus or the other, and thus they spend equal time between nuclei or not. These electrons neither contribute nor detract from bond strength.

VALENCE BOND AND MOLECULAR ORBITAL THEORY

In some respects valence bond theory is superior to molecular orbital theory. When applied to the simplest two-electron molecule, H_2, valence bond theory, even at the simplest Heitler-London approach, gives a much closer approximation to the bond energy, and it provides a much more accurate representation of the behaviour of the electrons as chemical bonds are formed and broken. In contrast simple molecular orbital theory predicts that the hydrogen molecule dissociates into a linear superposition of hydrogen atoms and positive and negative hydrogen ions, a completely unphysical result. This explains in part why the curve of total energy against interatomic distance for the valence bond method lies above the curve for the molecular orbital method at all distances and most particularly so for large distances. This situation arises for all homonuclear diatomic molecules and is particularly a problem for F_2, where the minimum energy of the curve with molecular orbital theory is still higher in energy than the energy of two F atoms.

The concepts of hybridization are so versatile, and the variability in bonding in most organic compounds is so modest, that valence bond theory remains an integral part of the vocabulary of organic chemistry. However, the work of Friedrich Hund, Robert Mulliken, and Gerhard Herzberg showed that molecular orbital theory provided a more appropriate description of the spectroscopic, ionization and magnetic properties of molecules. The deficiencies of valence bond theory became apparent when hypervalent molecules (e.g. PF_5) were explained without the use of d orbitals that were crucial to the bonding hybridisation scheme proposed for such molecules by Pauling. Metal complexes and electron deficient compounds (e.g. diborane) also appeared to be well described by molecular orbital theory, although valence bond descriptions have been made.

In the 1930s the two methods strongly competed until it was realised that they are both approximations to a better theory. If we take the simple valence bond structure and mix in all possible covalent and ionic structures arising from a particular set of atomic orbitals, we reach what is called the full configuration interaction wave function. If we take the simple molecular orbital description of the ground state and combine that function with the functions describing all possible excited states using unoccupied orbitals arising from the same set of atomic orbitals, we also reach the full configuration interaction wavefunction. It can be then seen that the simple molecular orbital approach gives too much weight to the ionic structures, while the simple valence bond approach gives too little. This can also be described as saying that the molecular orbital approach is too delocalised, while the valence bond approach is too localised.

The two approaches are now regarded as complementary, each providing its own insights into the problem of chemical bonding. Modern calculations in quantum chemistry usually start from (but ultimately go far beyond) a molecular orbital rather than a valence bond approach, not because of any intrinsic superiority in the former but rather because the MO approach is more readily adapted to numerical computations. However better valence bond programs are now available.

BONDS IN CHEMICAL FORMULAS

The 3-dimensionality of atoms and molecules makes it difficult to use a single technique for indicating orbitals and bonds. In molecular formulas the chemical bonds (binding orbitals) between atoms are indicated by various different methods according to the type of discussion. Sometimes, they are completely neglected. For example, in organic chemistry chemists are sometimes concerned only with the functional groups of the molecule.

Thus, the molecular formula of ethanol (a compound in alcoholic beverages) may be written in a paper in conformational, 3-dimensional, full 2-dimensional (indicating every bond with no 3-dimensional directions), compressed 2-dimensional (CH_3–CH_2–OH), separating the functional group from another part of the molecule (C_2H_5OH), or by its atomic constituents (C_2H_6O), according to what is discussed. Sometimes, even the non-bonding valence shell electrons (with the 2-dimensionalized approximate directions) are marked, i.e. for elemental carbon .'C'. Some chemists may also mark the respective orbitals, i.e. the hypothetical ethene"4 anion (\/C=C/ \ "4) indicating the possibility of bond formation.

CHEMICAL BONDS

These chemical bonds are intramolecular forces, which

hold atoms together in molecules. In the simplistic localized view of bonding, the number of electrons participating in a bond (or located in a bonding orbital) is typically multiples of two, four, or six, respectively. Even numbers are common because electrons enjoy lower energy states, if paired. Substantially more advanced bonding theories have shown that bond strength is not always a whole number, depending on the distribution of electrons to each atom involved in a bond.

For example, the carbons in benzene are connected to each other with about 1.5 bonds, and the two atoms in nitric oxide NO, are connected with about 2.5 bonds. Quadruple bonds are also well known. The type of strong bond depends on the difference in electronegativity and the distribution of the electron orbital paths available to the atoms that are bonded. The larger the difference in electronegativity, the more an electron is attracted to a particular atom involved in the bond, and the more "ionic" properties the bond is said to have ("ionic" means the bond electrons are unequally shared). The smaller the difference in electronegativity, the more covalent properties (full sharing) the bond has.

Molecular Mass Spectrometry

Mass spectrometry provides a powerful analytical tool for probing molecules for their bonding structure. Mass spectrometers are instruments that fragment and ionize "parent" molecules with electron beams and analyse the products with electric and magnetic fields and detectors. According to the laws of electromagnetism, charged particles passing through electromagnetic fields follow trajectories determined by their mass to charge ratio. Thus fragments with different masses and/or different charges are separated by the external fields, except in the case where the mass to charge ratio happens to be the same. E.g. O+ (m/q = 16/1) will be distinguished from O_2+ (m/q = 32/1) but not from O_2+ 2 (m/q = 32/2 = 16/1). Relative amounts of parents and fragments measured by the detector reflect relative abundances in the sample.

Mass spectra patterns give clues to the bonding of molecules. The first direct evidence that hydrogen is diatomic came in the early 1920's from a mass spectrometric analysis that showed two lines in the mass spectrum of elemental hydrogen, one with twice the mass as the other.

More complicated molecules can have more complicated mass spectra, and computers are used to assist in the analysis. Complicating features include multiple combinations of possible "daughter" fragments, multiple ionizations of the parent and fragments, and multiple isotopes.

If the molecules have a chance to react before analysis, fragments can recombine in new ways to confuse the analysis. A good way to analyse or predict a mass spectrum tabulates the possible combinations of parent and fragment masses, using isotope distributions to determine the spectral lines.

ATOMIC STRUCTURE AND BONDING

Atoms are composed of electrons, protons, and neutrons. Electron and protons are negative and positive charges of the same magnitude, 1.6 × 10-19 Coulombs. The mass of the electron is negligible with respect to those of the proton and the neutron, which form the nucleus of the atom.

The unit of mass is an atomic mass unit (amu) = 1.66 × 10-27 kg, and equals 1/12 the mass of a carbon atom. The Carbon nucleus has Z=6, and A=6, where Z is the number of protons, and A the number of neutrons. Neutrons and protons have very similar masses, roughly equal to 1 amu. A neutral atom has the same number of electrons and protons, Z.

A mole is the amount of matter that has a mass in grams equal to the atomic mass in amu of the atoms. Thus, a mole of carbon has a mass of 12 grams. The number of atoms in a mole is called the Avogadro number, Nav = 6.023×10^{23}. Note that Nav = 1 gram/1 amu.

Calculating n, the number of atoms per cm^3 in a piece of material of density (g/cm^3).

$$n = Nav \times / M$$

where M is the atomic mass in amu (grams per mol). Thus, for graphite (carbon) with a density ? = 1.8 g/cm3, M =12, we get 6×10^{23} atoms/mol $\times$ 1.8 g/cm^3/ 12 g/mol) = 9×10^{22} C/cm^3. For a molecular solid like ice, one uses the molecular mass, $M(H_2O) = 18$. With a density of 1 g/cm^3, one obtains n = 3.3×10^{22} H_2O/cm^3. Note that since the water molecule contains 3 atoms, this is equivalent to 9.9×10^{22} atoms/cm^3.

Most solids have atomic densities around 6 × 1022 atoms/ cm3. The cube root of that number gives the number of atoms per centimeter, about 39 million. The mean distance between atoms is the inverse of that, or 0.25 nm. This is an important number that gives the scale of atomic structures in solids.

ELECTRONS IN ATOMS

The forces in the atom are repulsions between electrons and attraction between electrons and protons. The neutrons play no significant role. Thus, Z is what characterizes the atom. The electrons form a cloud around the neutron, of radius of 0.05 – 2 nanometers.

Electrons do not move in circular orbits, as in popular drawings, but in 'fuzzy' orbits. We cannot tell how it moves, but only say what is the probability of finding it at some distance from the nucleus. According to quantum mechanics, only certain orbits are allowed (thus, the idea of a mini planetary system is not correct). The orbits are identified by a principal quantum number n, which can be related to the size, n = 0 is the smallest; n = 1, 2 .. are larger. (They are "quantized" or discrete, being specified by integers). The angular momentum l is quantized, and so is the projection in a specific direction m. The structure of the atom is determined by the Pauli exclusion principle, only two electrons can be placed in an orbit with a given n, l, m – one for each spin. Table in the textbook gives the number of electrons in each shell (given by n) and subshells (given by l).

PERIODIC TABLE

Elements are categorized by placing them in the periodic table. Elements in a column share similar properties. The noble gases have closed shells, and so they do not gain or lose electrons near another atom. Alkalis can easily lose an electron and become a closed shell; halogens can easily gain one to form a negative ion, again with a closed shell. The propensity to form closed shells occurs in molecules, when they share electrons to close a molecular shell. Examples are H_2, N_2, and NaCl. The ability to gain or lose electrons is termed electronegativity or electropositivity, an important factor in ionic bonds.

BONDING FORCES AND ENERGIES

The Coulomb forces are simple: attractive between electrons and nuclei, repulsive between electrons and between nuclei. The force between atoms is given by a sum of all the individual forces, and the fact that the electrons are located outside the atom and the nucleus in the centre.

When two atoms come very close, the force between them is always repulsive, because the electrons stay outside and the nuclei repel each other. Unless both atoms are ions of the same charge (e.g., both negative) the forces between atoms is always attractive at large internuclear distances r. Since the force is repulsive at small r, and attractive at small r, there is a distance at which the force is zero. This is the equilibrium distance at which the atoms prefer to stay.

The interaction energy is the potential energy between the atoms. It is negative if the atoms are bound and positive if they can move away from each other. The interaction energy is the integral of the force over the separation distance, so these two quantities are directly related. The interaction energy is a minimum at the equilibrium position. This value of the energy is called the bond energy, and is the energy needed to separate completely to infinity (the work that needs to be done to overcome the attractive force.) The strongest the bond energy, the hardest is to move the atoms, for instance the hardest it is to melt the solid, or to evaporate its atoms.

PRIMARY INTERATOMIC BONDSIONIC BONDING

This is the bond when one of the atoms is negative (has an extra electron) and another is positive (has lost an electron). Then there is a strong, direct Coulomb attraction. An example is NaCl. In the molecule, there are more electrons around Cl, forming Cl^- and less around Na, forming Na^+. Ionic bonds are the strongest bonds.

In real solids, ionic bonding is usually combined with covalent bonding. In this case, the fractional ionic bonding is defined as %ionic = 100 × [1 – exp(-0.25 (XA – XB)2], where XA and XB are the electronegativities of the two atoms, A and B, forming the molecule.

Electronic Theory of Chemical Bonding

Nowadays it is recognized that the chemical bond is the most basic and essential feature of molecules. Understanding the properties of the interatomic bonds in molecules is essential to understanding the physical structure and chemical reactivities of molecular matter. Chemical bonds result from the electrostatic (Coulombic) interactions of the electrons and nuclei of atoms.

Basically, the outer-shell electrons on each atom in a molecule can be thought to be attracted to the nuclei of adjacent atoms as well as their own nuclei. Figure diagrams the electrostatic interactions for a classical particle model of the dihydrogen molecule, the simplest neutral polyatomic molecule. Actually this model, although useful, is an oversimplification on several counts; as was found for atoms, electrons in the presence of nuclei do not behave as particles and classical models fall short of accurate descriptions of electronic structure.

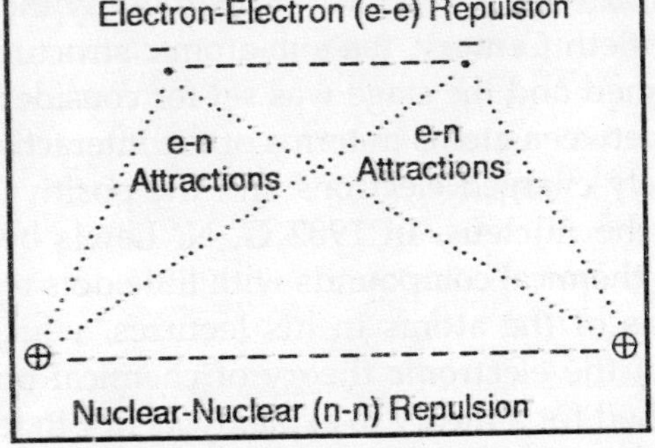

Fig. Electrostatic Interactions in Dihydrogen

The wave mechanical model of electronic structure is the appropriate description for multiple electron systems, and applies to molecules as well as atoms. The nature of the chemical bond is revealed in the solutions to Schr"odinger's equation for molecules.

As in the case of atoms, accurate solutions obtained for simple molecules are used to justify established bonding heuristics, as well as generate new molecular bonding principles. A major feature that emerges in investigating chemical bonds through wave mechanics is electrons prefer to be paired. This fact explains the stability of the simplest neutral molecule, H_2 (with 2 paired electrons) with respect to H atoms, as well as the instability of He_2 with respect to He atoms (each having 2 paired electrons). The stability of paired electrons and filled subshell electronic configurations prove to be most useful general principles for explaining a variety of features of molecules.

Before delving further into the nature of the chemical bond as elucidated by the wave mechanics, we will explore a widely used powerful pre-wave mechanical bonding heuristic invented by G. N. Lewis early in the beginning of the Twentieth Century. Lewis considered electrons to be classical particles, which he represented graphically by dots. Pairs of electrons between atoms he represented with lines to represent bonds connecting the atoms together.

Lewis's contribution consisted of a simple method of counting the dots to predict chemical formulas, molecular geometry, and various chemical properties. By the beginning of the Twentieth Century, the sub-atomic structure of atoms was established and the stage was set for considering atomic attractions between atoms in terms of the interactive forces of the negatively-charged electrons and the positively-charged protons of the nucleus. In 1902 G. N. Lewis began using formulas of chemical compounds with little dots representing the electrons of the atoms in his lectures, which was the beginning of the electronic theory of chemical bonding. As Lewis searched for a theory to explain valency in terms of the electrons of an atom, he was led to the periodic table. We now know that columns of chemical families have similar properties because they have similar outer-shell electronic configurations, but that knowledge wasn't available at the beginning of the century.

Lewis reasoned that families of elements have similar properties because they have similar numbers of electrons participating in the bonding. This led to dividing the electrons of an atom into two categories, now known as inner shell, or core electrons and outer shell or valence electrons. There may be multiple inner shells of electrons, but the outer shell valence electrons are the ones closest to neighboring nuclei and therefore dominate in the chemical bonding. A chemically stable (inert) noble gas is found at the end of each row of the periodic table.

The fact that each new noble gas has added at least eight more electrons, suggests some connection between eight electrons and stability. Atoms to the left and right of the noble gases have fewer than eight electrons, are less stable, and are hence chemically reactive. Lewis devised the octet rule to describe the tendency of atoms to gain stability by achieving an octet of outer shell, or valence electrons.

However, there are exceptions. Elements in the first three columns of the periodic table may form suboctets (less than 8 electrons) and elements beyond the second row of the periodic table can form superoctets (more than 8 electrons). Hydrogen is a notable exception, requiring merely a duet or pair of electrons to acquire the stability of the noble gas helium. The duet/octet heuristic, or "rule" states that atoms like to be noble (or have noble gas electronic configurations), or, in modern terms, filled subshells in electronic configurations are particularly stable.

The octet heuristic was successful in describing the bonding in both ionic compounds (containing charged atoms or molecules) and covalent molecules (combinations of non-charged atoms) in terms of the valence electrons of atoms. In addition, the stability of paired electrons for the first time offered a rationalization4 of the stability of homonuclear diatomic gases, such as dihydrogen and dioxygen.

LEWIS STRUCTURES

Covalent pairs of electrons Lewis called bond pairs. Atoms with valence greater than unity could possibly share more than one pair of electrons with a neighboring atom, producing stronger multiple bonds.5 Electrons left over after all the valences of the atoms were satisfied, Lewis classified as lone pair electrons.

The identification of different classes of electrons in molecules proved very powerful in explaining their structures and reactivities. Molecular formulas which show the bonding and free electrons are called Lewis structure formulas. Lewis structures based on the duet/octet heuristic provide a simplified basis for explaining the properties of molecules.

ELECTRON CORRELATIONS IN SOLIDS

The variety of phenomena associated with the electronic structure of crystalline matter has been the subject of theoretical investigation for a long time. Historically, the starting impetus was given by the challenge of accounting for some peculiar properties of the metallic state, such as the excellent heat and electric conductivity. The quantum theory of electrons moving in a periodic lattice developed in the 1920's by Sommerfeld, Bloch, Wilson and others, for example, was indeed primarily aimed at understanding the electronic transport properties of metals.

This theory led to the band model of the crystalline solid. It was realized that within the band model the configuration of the valence electrons can explain the division of perfect crystals at low temperatures into metals and insulators. If the valence electrons exactly fill one or more energy bands, leaving others empty, the crystal is an insulator. The difference in energy between the top of the highest occupied band and the bottom of the lowest empty band is known as the band gap. In the other case, when partially filled bands exist, the solid has metallic properties.

The band description of solids rests, as the simple orbital theories of atoms and molecules, on the independent electron approximation. Each electron experiences the periodic field due to the atomic nuclei and an effective field obtained by averaging over the positions of the remaining electrons. In spite of this approximation band theory has been very successful in predicting and understanding the properties of real materials, including elemental metals and many (simple) covalent and ionic compounds. It has been noticed, however, already in the 1930's that for some solids the one-electron band theory is inadequate. These are systems where, due to the competition between delocalization, band-like, effects and strong (atomic-like) electron – electron interactions, the independent electron approximation breaks down. Typical examples are the 3d -metal compounds.

In these compounds strong repulsive interactions may result in electron localization and the opening of a conductivity gap even though the d-"bands"are not completely filled. The presence of bound, unpaired, 3d electrons further gives rise to local magnetic moments and very rich magnetic properties. Serious attempts to characterize and understand the electronic structure of such materials began in the late 1950's. The main motivations at the time came from the issue of the Mott metal – insulator transition, the problem of magnetic ordering, and the problem of itinerant ferromagnetism.

More recently, phenomena like heavy fermion behaviour, high-temperature superconductivity, colossal magnetoresistance, and spin-Peierls phase transitions have revived interest in these systems. Nevertheless, although considerable effort has been put into the field, many of the transition metal materials are poorly understood. The proper treatment of various competing physical effects, like electron localization as a result of strong electron – electron interactions and band-like behaviour as a result of orbital overlap and translational symmetry, remains one of the most difficult problems in theoretical solid state physics.

Wave-function based correlated quantum chemical methods have been successfully applied for calculating molecular electronic structures. A very attractive feature of the quantum chemical approach is the ability to address the electron correlation problem in a systematic and controlled manner. However, due to the high computational cost of these methods, inclusion of electron correlation effects in the study of crystalline solids did not seem feasible for a long time. Algorithms able to treat electron correlation explicitly in extended periodic systems have been developed only recently, either within the many-body second-order perturbation theory (MBPT2) or within the configuration interaction (CI) formalism. Still, calculations have been carried out for rather simple closed-shell systems only, like crystalline LiH and polymers.

Currently the most widely used first-principles method in solid state applications is the density functional (DF) theory within the local density approximation (LDA) (for a review). In the DF formalism, one avoids construction of the many-body wave-function and instead computes specific properties of the system from the ground-state electronic charge density. Recently, the study of the electronic response to external perturbations made possible calculation of low-energy excitations such as phonon dispersions, optical spectra and magnon dispersions.

The DF-based methods approximately include electron correlation and usually give accurate estimates for a number of physical properties — calculated crystal structures, lattice parameters, elastic constants, dielectric functions and phonon frequencies are within a few per cent of the experimental values. Notable exception is the underestimation of the band gap. For strongly correlated 3d -electron systems like CoO and La_2CuO_4, DF/LDA calculations fail actually in predicting the correct insulating ground-state.

A different approach to the problem of strongly interacting electrons in solids is the model Hamiltonian method, where many-body effects are incorporated in some parameterized form. Two of the simplest models are the Hubbard Hamiltonian widely used in a qualitative or a semi-quantitative way for studying the interplay between electron delocalization and electron repulsion effects, and the Anderson model of a magnetic impurity coupled to a conduction band. Model Hamiltonian investigations can often describe tendencies and reveal the basic mechanisms of various physical phenomena. At the same time the model Hamiltonian approach is rather limited in its ability to make quantitative predictions, since even the simple Hubbard model can be solved exactly only in the one-dimensional case. In dealing with the two- and three-dimensional models numerical techniques are usually employed, based on quantum Monte Carlo methods or on diagonalizing the Hamiltonian for a cluster of atoms. However, the quantum Monte Carlo methods suffer from the so-called sign problem, while the exact diagonalization is limited by the exponential growth of the computational effort with the cluster size.

Local cluster (or atomic) impurity-like models have been used to describe various electronic properties of strongly correlated 3d- and 4f-metal systems for a long time, in combination with model Hamiltonians or with wave-function based quantum chemical calculations. Within the ab initio wave-function based approach the cluster is embedded in some effective potential that accounts for the crystal Madelung field and for short-range Pauli and exchange interactions due to the finite charge distribution of the nearest neighbors, respectively. The cluster contains one or more metal ions and the adjacent ligands. Proper treatment of the many-body problem, by configuration interaction techniques or second-order perturbation theory gives in several insulating transition metal materials accurate estimates and predictions for properties like core level spectra d – d excitations ligand to metal charge transfer effects and Heisenberg spin interactions.

In the study of the local many-body physics in solids a step forward consists in solving the problem of a (single- or multi-site) quantum impurity embedded in an effective medium self-consistently. This is the essential idea behind the dynamical mean field theory (DMFT), developed during the last decade (for a review). The DMFT reduces a lattice model to an Anderson impurity subject to a self-consistently determined "hybridization" with an electron "bath" representing the crystalline host.

Its structure is that of a closed set of equations relating

the "correlated"single-site impurity Green function to the crystal Green function evaluated at the impurity site. The crystal Green function can be calculated using standard DF/ LDA band structure methods. Within such a LDA+DMFT scheme the DMFT correction to the LDA input results in a change of the electronic charge density. This updated charge density is used in a new LDA+DMFT loop and the cycle is repeated until convergence is reached. Results of such calculations were able, for example, to describe the finite-temperature magnetic properties of iron and nickel and the phase diagram of plutonium. Efforts to extend the single-site DMFT to include multi-site, short-range correlations led to the dynamical cluster approximation (DCA) and the cellular dynamical field theory (CDMFT).

BAND STRUCTURE OF ELECTRONIC

The electronic band structure (or simply band structure) of a solid describes ranges of energy that an electron is "forbidden" or "allowed" to have. It is due to the diffraction of the quantum mechanical electron waves in the periodic crystal lattice. The band structure of a material determines several characteristics, in particular its electronic and optical properties.

BANDS OCCUR IN MATERIALS

The electrons of a single free-standing atom occupy atomic orbitals, which form a discrete set of energy levels. If several atoms are brought together into a molecule, their atomic orbitals split, as in a coupled oscillation. This produces a number of molecular orbitals proportional to the number of atoms. When a large number of atoms (of order 1020 or more) are brought together to form a solid, the number of orbitals becomes exceedingly large, and the difference in energy between them becomes very small, so the levels may be considered to form continuous bands of energy rather than the discrete energy levels of the atoms in isolation. However, some intervals of energy contain no orbitals, no matter how many atoms are aggregated, forming band gaps.

Within an energy band, energy levels are so numerous as to be a near continuum. First, the separation between energy levels in a solid is comparable with the energy that electrons constantly exchange with phonons (atomic vibrations). Second, it is comparable with the energy uncertainty due to the Heisenberg uncertainty principle, for reasonably long intervals of time. As a result, the separation between energy levels is of no consequence.

Basic Concepts

Any solid has a large number of bands. In theory, it can be said to have infinitely many bands. However, all but a few lie at energies so high that any electron that reaches those energies escapes from the solid.

These bands are usually disregarded.Bands have different widths, based upon the properties of the atomic orbitals from which they arise. Also, allowed bands may overlap, producing (for practical purposes) a single large band.Metals contain a band that is partly empty and partly filled regardless of temperature.

Therefore they have very high conductivity.The lowermost, almost fully occupied band in an insulator or semiconductor is called the valence band by analogy with the valence electrons of individual atoms. The uppermost, almost unoccupied band is called the conduction band because only when electrons are excited to the conduction band can current flow in these materials.

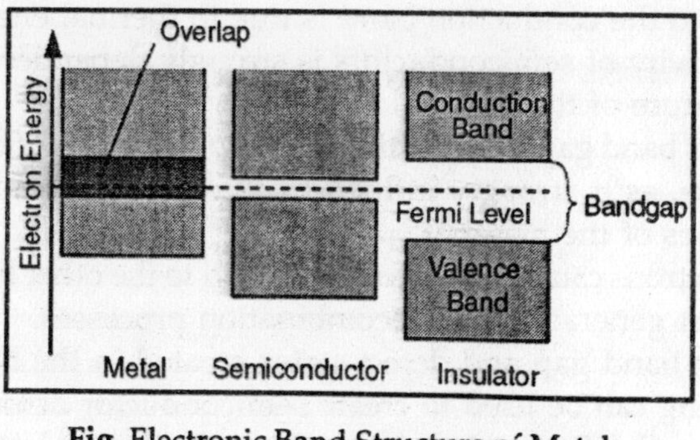

Fig. Electronic Band Structure of Metals.

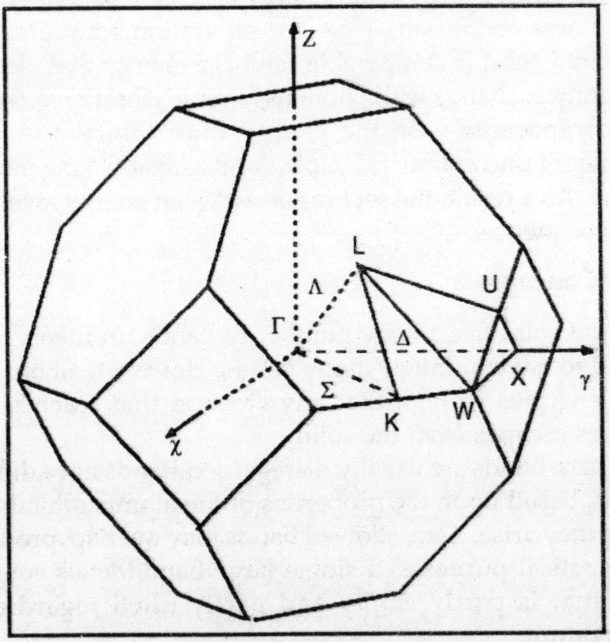

Fig. First Brillouin Zone of FCC Lattice

The difference between insulators and semiconductors is only that the forbidden band gap between the valence band and conduction band is larger in an insulator, so that fewer electrons are found there and the electrical conductivity is lower.

Because one of the main mechanisms for electrons to be excited to the conduction band is due to thermal energy, the conductivity of semiconductors is strongly dependent on the temperature of the material.

This band gap is one of the most useful aspects of the band structure, as it strongly influences the electrical and optical properties of the material.

Electrons can transfer from one band to the other by means of carrier generation and recombination processes.

The band gap and defect states created in the band gap by doping can be used to create semiconductor devices such as solar cells, diodes, transistors, laser diodes, and others.

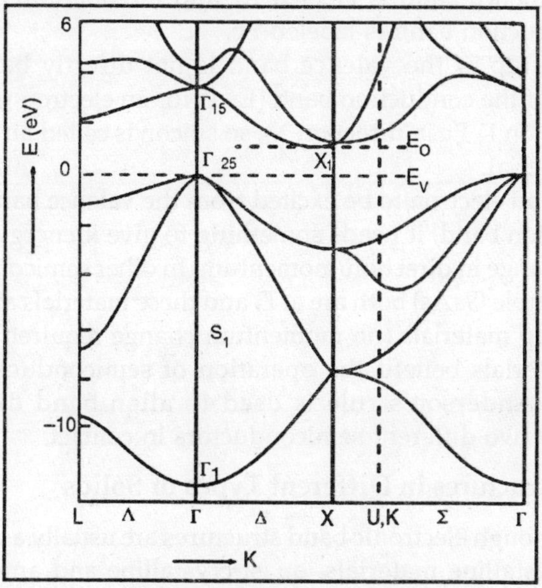

Fig. Band Structure in Silicon

Symmetry

A more complete view of the band structure takes into account the periodic nature of a crystal lattice using the symmetry operations that form a space group. The Schrödinger equation is solved for the crystal, which has Bloch waves as solutions:

$$\psi_{nk}(r) = e^{ik.r}u_{nk}(r),$$

where *k* is called the wavevector, and is related to the direction of motion of the electron in the crystal, and n is the band index, which simply numbers the energy bands. The wavevector k takes on values within the Brillouin zone (BZ) corresponding to the crystal lattice, and particular directions/points in the BZ are assigned conventional names like Γ Δ Λ Σ etc. These directions are shown for the face-centered cubic lattice geometry in.

The available energies for the electron also depend upon k, as shown in Figure for silicon in the more complex energy band diagram at the right. In this diagram the topmost energy of the valence band is labeled Ev and the bottom energy in the conduction band is labeled Ec.

The top of the valence band is not directly below the bottom of the conduction band (Ev is for an electron traveling in direction Γ, Ec in direction X), so silicon is called an indirect gap material.

For an electron to be excited from the valence band to the conduction band, it needs something to give it energy Ec "Ev and a change in direction/momentum. In other semiconductors (for example GaAs) both are at Γ, and these materials are called direct gap materials (no momentum change required). Direct gap materials benefit the operation of semiconductor laser diodes. Anderson's rule is used to align band diagrams between two different semiconductors in contact.

Band Structures in Different Types of Solids

Although electronic band structures are usually associated with crystalline materials, quasi-crystalline and amorphous solids may also exhibit band structures. However, the periodic nature and symmetrical properties of crystalline materials makes it much easier to examine the band structures of these materials theoretically. In addition, the well-defined symmetry axes of crystalline materials makes it possible to determine the dispersion relationship between the momentum (a 3-dimension vector quantity) and energy of a material. As a result, virtually all of the existing theoretical work on the electronic band structure of solids has focused on crystalline materials.

Density of States

While the density of energy states in a band could be very large for some materials, it may not be uniform. It approaches zero at the band boundaries, and is generally highest near the middle of a band. The density of states for the free electron model in three dimensions is given by,

$$D(\hat{I}) = \frac{V}{2\pi^2}\left(\frac{2m}{\hbar^2}\right)^{3/2} \in^{1/2}$$

Filling of Bands

Although the number of states in all of the bands is effectively infinite, in an uncharged material the number of electrons is equal only to the number of protons in the atoms of the material.

Therefore not all of the states are occupied by electrons ("filled") at any time. The likelihood of any particular state being filled at any temperature is given by the Fermi-Dirac statistics. The probability is given by the following:

$$f(E) = \frac{1}{1 + e^{\frac{E - E_F}{k_B T}}}$$

where:

k_B is Boltzmann's constant,

T is the temperature,

E_F is the Fermi energy (or 'Fermi level').

The Fermi level naturally is the level at which the electrons and protons are balanced.

At $T = 0$, the distribution is a simple step function:

$$f(E) = \begin{cases} 1 & \text{if } 0 < E \leq E_F \\ 0 & \text{if } E_F < E \end{cases}$$

At nonzero temperatures, the step "smooths out", so that an appreciable number of states below the Fermi level are empty, and some states above the Fermi level are filled.

BAND STRUCTURE OF CRYSTALS

BRILLOUIN ZONE

Because electron momentum is the reciprocal of space, the dispersion relation between the energy and momentum of electrons can best be described in reciprocal space. It turns out that for crystalline structures, the dispersion relation of the electrons is periodic, and that the Brillouin zone is the smallest repeating space within this periodic structure. For an infinitely large crystal, if the dispersion relation for an electron is defined throughout the Brillouin zone, then it is defined throughout the entire reciprocal space.

THEORY OF BAND STRUCTURES

The ansatz is the special case of electron waves in a periodic crystal lattice using Bloch waves as treated generally in the dynamical theory of diffraction. Every crystal is a periodic structure which can be characterized by a Bravais lattice, and for each Bravais lattice we can determine the reciprocal lattice, which encapsulates the periodicity in a set of three reciprocal lattice vectors (b_1, b_2, b_3). Now, any periodic potential $V(r)$ which shares the same periodicity as the direct lattice can be expanded out as a Fourier series whose only non-vanishing components are those associated with the reciprocal lattice vectors.

So the expansion can be written as:

$$V(r) = \sum_K V_k e^{iK.r}$$

where $K = m_1 b_1 + m_2 b_2 + m_3 b_3$ for any set of integers (m_1, m_2, m_3).

From this theory, an attempt can be made to predict the band structure of a particular material, however most ab initio methods for electronic structure calculations fail to predict the observed band gap.

Nearly-free Electron Approximation

In the nearly-free electron approximation in solid state physics interactions between electrons are completely ignored. This approximation allows use of Bloch's Theorem which states that electrons in a periodic potential have wavefunctions and energies which are periodic in wavevector up to a constant phase shift between neighboring reciprocal lattice vectors.

The consequences of periodicity are described mathematically by the Bloch wavefunction:

$$Y_{n,k}(r) = e^{ik.r} u_n(r)$$

where the function $u_n(r)$ is periodic over the crystal lattice, that is,

$$u_n(r) = u_n(r - R).$$

Here index n refers to the n-th energy band, wavevector k is related to the direction of motion of the electron, r is position in the crystal, and R is location of an atomic site.

Tight-binding Model

The opposite extreme to the nearly-free electron approximation assumes the electrons in the crystal behave much like an assembly of constituent atoms. This tight-binding model assumes the solution to the time-independent single electron Schrödinger equation Ø is well approximated by a linear combination of atomic orbitals $\psi n(r)$.

$$\Upsilon(r) = \sum b_{n,R}\psi_n(r - R),$$

where the coefficients $b_{n,R}$ are selected to give the best approximate solution of this form. Index n refers to an atomic energy level and R refers to an atomic site. A more accurate approach using this idea employs Wannier functions, defined by,

$$a_n(r - R) = \frac{V_C}{(2\pi)^3} \int_{BZ} dk e^{-ik.(R-r)} u_{nk};$$

in which u_{nk} is the periodic part of the Bloch wave and the integral is over the Brillouin zone. Here index n refers to the n-th energy band in the crystal. The Wannier functions are localized near atomic sites, like atomic orbitals, but being defined in terms of Bloch functions they are accurately related to solutions based upon the crystal potential. Wannier functions on different atomic sites R are orthogonal. The Wannier functions can be used to form the Schrödinger solution for the n-th energy band as:

$$\Upsilon_{n,k}(r) = \sum_R e^{-ik.(R-r)} a_n(r - R)$$

Ab initio Density-functional Theory

In present days physics literature, the large majority of the electronic structures and band plots is calculated using the density-functional theory (DFT) which is not a model but rather an ab initio theory, i.e. a microscopic first-principle theory of condensed matter physics that tries to cope with the electron-electron many-body problem via the introduction of an exchange-correlation term in the functional of the electronic density. DFT calculated bands are found in many cases in agreement with experimental measured bands, for example by angle-resolved photoemission spectroscopy (ARPES). In particular, the band shape seems well reproduced by DFT. But also there are systematic errors of DFT bands with respect to the experiment. In particular, DFT seems to underestimate systematically by a 30-40% the band gap in insulators and semiconductors.

It must be said that DFT is in principle an exact theory to reproduce and predict ground state properties (e.g. the total energy, the atomic structure, etc.). However DFT is not a theory to address excited state properties, such as the band plot of a solid that represents the excitation energies of electrons injected or removed from the system. What in literature is quoted as a DFT band plot is a representation of the DFT Kohn-Sham energies, that is the energies of a fictive non-interacting system, the Kohn-Sham system, which has no physical interpretation at all.

The Kohn-Sham electronic structure must not be confused with the real, quasiparticle electronic structure of a system, and there is no Koopman's theorem holding for Kohn-Sham energies, like on the other hand for Hartree-Fock energies that can be truly considered as an approximation for quasiparticle energies. Hence in principle DFT is not a band theory, not a theory suitable to calculate bands and band-plots.

GREEN'S FUNCTION

To calculate the bands including electron-electron interaction many-body effects, one can resort to so called Green's function methods. Indeed, the knowledge of the Green's function of a system provides both ground (the total energy) and also excited state observables of the system. The poles of the Green's function are the quasiparticle energies, the bands of a solid. The Green's function can be calculated by solving the Dyson equation once the self-energy of the system is known. For real systems like solids, the self-energy is a very complex quantity and usually approximations are needed to solve the problem.

One of such approximations is the GW approximation,

so called from the mathematical form the self-energy takes as product Ó = GW of the Green's function G and the dynamically screened interaction W.

This approach is more pertinent to address the calculation of band plots (and also quantities beyond, such as the spectral function) and can be also formulated in a completely ab initio way. The GW approximation seems to provide band gaps of insulators and semiconductors in agreement with the experiment and hence to correct the systematic DFT underestimation.

Mott Insulators

Although the nearly-free electron approximation is able to describe many properties of electron band structures, one consequence of this theory is that it predicts the same number of electrons in each unit cell. If the number of electrons is odd, we would then expect that there is an unpaired electron in each unit cell, and thus that the valence band is not fully occupied, making the material a conductor.

However, materials such as CoO that have an odd number of electrons per unit cell are insulators, in direct conflict with this result. This kind of material is known as a Mott insulator, and requires inclusion of detailed electron-electron interactions (treated only as an averaged effect on the crystal potential in band theory) to explain the discrepancy. The Hubbard model is an approximate theory that can include these interactions.

ENERGY BANDS

The electronic conductivity of a material is determined by the properties of its constituent atoms or molecules, and by the manner in which they are arranged in the lattice. A solid-state model that relates electronic processes to valance and conduction energy bands. The valance band consists of electrons that, because they have relatively low energy, are associated with individual atoms or molecules: the conduction band contains more energetic electrons that are free to move throughout the material in response to applied electromagnetic energy.

The number and mobility of conduction electrons determines the electronic conductivity of a material. If the valance and conduction bands are separated by a small gap, then, at typical temperatures, thermal activity will deplete the valance band and populate the conduction band; such a material is a conductor. If the bands are widely separated in energy, the conduction band will be vacant and the material will be an insulator. A semiconductor is a material whose band structure falls between that of a conductor and an insulator-it can be an insulator at one temperature and a conductor at a higher temperature. Semiconductors can contain impurity atoms whose energy states lie within the gap between the valance and conduction bands; such impurities strongly affect conductivity by donating or accepting electrons.

An important consequence of the existence of energy bands is that they permit electronic processes in one region of a material to affect not only the immediate area, but also the entire structure. Szent-Gyorgyi proposed that common energy levels existed over relatively large dimensions in biological structures, possibly with the cell wall itself as the boundary. Evans and Gergely calculated the band gap in hydrogen-bonded models of biopolymers and showed that it would be so large that the biopolymers would behave electrically as insulators. However, if impurity atoms were present, they could donate an electron to the conduction band, or remove one from the valance band, leading to mobile conduction electrons or mobile "holes" in the valance band. Szent-Gyorgyi postulated that these electronic processes within the energy bands-electron mobility in the conduction band and charge transfer in the valance band-could give rise to biological phenomena and, indeed, to life itself.

Fig. A Large Protein Molecule Contains Many Electron Pairs

For ordinary materials the question of their band structure could be resolved by a coordinated series of X-ray, chemical, and electrondynamics studies. But biological tissue is inhomogeneous and impure, and suitable techniques for carrying out many of the necessary studies on such materials have not yet been developed. Perhaps the most significant problem for the experimentalist is that posed by the universal presence of water in tissue.

It is well established that the electrical conductivity of tissue increases sharply with water content. However, the nature of electrical conduction in tissue under physiological conditions of temperature and moisture-the relative contribution of electronic, protonic, and ionic processes-has not been established despite more than 30 years of study. Thus, no clear picture of the band structure in tissue has emerged. Other important solid-state techniques that have been used to study the electronic property of biological tissue include electron paramagnetic resonance, and photoconductivity. Again, although the results are consistent with a common-energy-band model proposed by Szent-Gyorgyi, they do not establish it as correct.

Chapter 4

Phases Equilibrium

PROPERTIES OF SOLIDS

Solids have a number of observable properties in common: They are rigid, maintaining fixed shape and volume independent of container.

The molar volumes of solids are small. Solids are incompressible. Solids do not flow, and diffusion in solids is extremely slow.

Many solids occur as crystals: Well-defined shapes with structures suggesting an orderly internal arrangement of atoms, molecules, or ions. To rationalize these properties we need a simple model of the solid phase.

In designing such a model, we make use of two facts:

- Since the molar volume of a solid is much less than the molar volume of the gas phase of a substance, the molecules must be crowded together in the solid;
- The orderly shapes of crystals suggests an orderly arrangement of particles. A simple model, then, is that solids consist of tightly-packed orderly arrangements of particles.

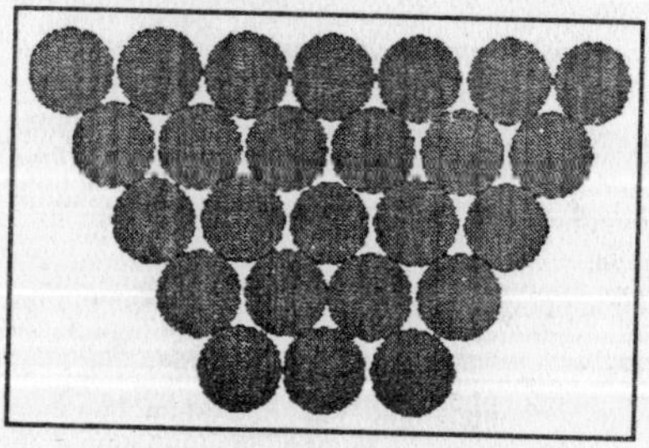

Fig. 2-dimensional of a Solid

This crowding allows very little free volume to the molecules, which strongly resist further compression. The solid is incompressible. It is also clear why diffusion is slow in solids, because each particle is locked in place by the surrounding ones. Only at the surface of the solid, where a particle has neighbors on only 3 sides, can diffusion occur. The particles are in constant motion, just as in the gas phase. A molecule is constantly jiggling in place, bumping into neighbors that hold it in a cage. Just as for gases, the kinetic energies of the particles (atoms, molecules, or ions) of the solid are distributed according to the Maxwell-Boltzmann distribution, with the kinetic energy of an average particle having the value 3kT/2.

The major difference between the motion in gas and solid is that in the solid, a particle is hemmed in and undergoes many more collisions per unit time than in the gas. The movement of a particle in the solid is much like a vibration, because the molecule stays in the same location. If the temperature of the solid is increased, the particles vibrate faster, but are still not free to move about because intermolecular forces from the neighbors hold them in place.

Finally, the arrangement of particles is very orderly, not at all chaotic like the gas. Each particle is surrounded by a perfect hexagon of other particles; and particles are arranged in rows that remain straight over long distances. It is easy to imagine that this orderly arrangement at the molecular level is manifested in beautiful crystalline shapes at the macroscopic level.

THE LATTICE

We will describe structures of crystalline solids in terms of the lattice. A lattice is a regular, repeating pattern of points in space. Wallpaper patterns are examples of 2-dimensional lattices. A wallpaper pattern and the equivalent lattice. The lattice is more fundamental than the particular wallpaper pattern shown, because wallpaper patterns using many different design units could be generated from the same lattice. The key feature of the lattice is the arrangement of points. A small collection of lattice points that contains all of the spatial information in the lattice pattern is called a unit cell. The unit cell represents the lattice, because the whole lattice can be reproduced from the unit cell by moving (translating) the unit cell in directions parallel to its sides by distances equal to the lengths of its sides.

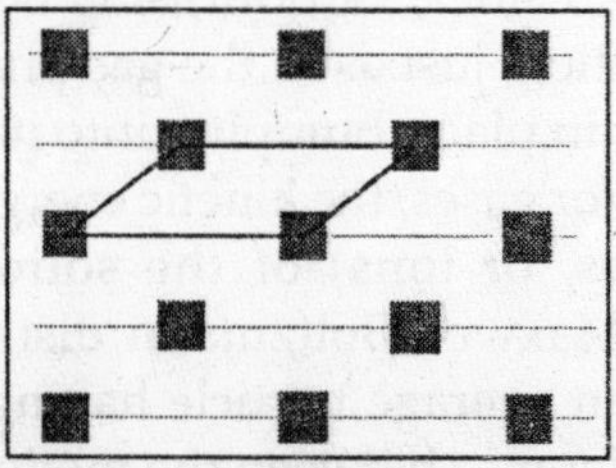

Fig. Unit Cell in the Lattice

Movement of this collection in directions parallel to its sides does indeed reproduce the entire lattice. However, the unit cell in Figure 7-2a is simpler than that in Figure 7-2b because it involves fewer points — one point at each of its 4 corners. A unit cell having lattice points at its corners only is called a primitive unit cell. The unit cell in Figure 7-2b is a non-primitive unit cell because it has an additional point at its centre. If it is possible to represent a lattice with a primitive unit cell, we will do so. Crystalline solids are represented by 3-dimensional lattices, one example of which is the simple cubic lattice. This is a primitive unit cell because it has points only at the corners. Movement of the unit cell by the distance, d, in directions parallel to its 3 dimensions reproduces the entire simple cubic lattice. The simple cubic lattice is analogous to an orderly arrangement of building blocks stacked face to face in 3-dimensions.

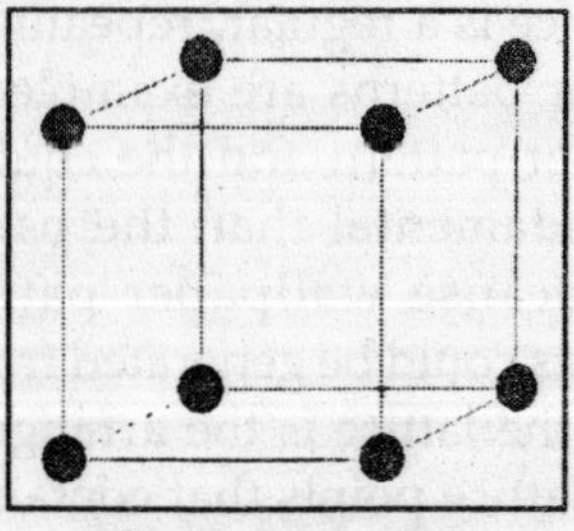

Fig. Cubic Lattice

It was shown in the 1800s by Bravais that there are only 7 shapes (called crystal types) that a unit cell can have and only 14 possible distributions of points (lattice types) within these 7 shapes.

The 7 shapes, characterized by the relative magnitudes of the repeat distances in three directions and the angles between these directions.

Table. The Bravais Crystal Types

System	Edge Lengths	Angles
Cubic	a = b = c	all 90°
Tetragonal	a = b not equal to c	all 90°
Orthorhombic	a not equal to b not equal to c	all 90°
Monoclinic (once-angled)	a not equal to b not equal to c	a = b = 90°, g not equal to 90°
triclinic (thrice-angled)	a not equal to b not equal to c to 90°	a not equal to b not equal to g not equal
Rhombohedral	a = b = c	= 90°
Hexagonal	a = b not equal to c	a = b = 90°; g = 120°

CLOSE PACKING

What is the most efficient way of arranging spheres in 3-dimensional space? We take "efficient" to mean economical in terms of space.

We begin by building one layer of spheres that are arranged as closely to one another as possible. Note that this arrangement is identical to that pictured earlier in Figure. Spheres are lined up in straight rows, with the spheres in one row fitting snugly into the indentations in adjacent rows.

Each sphere is surrounded by a perfect hexagon of spheres, all touching it. There are numerous gaps, or holes, in the layer, of triangular shape.

For descriptive purposes here, we label these holes "u" if the triangle has a single vertex pointing up, or "d" if the triangle has vertex down.

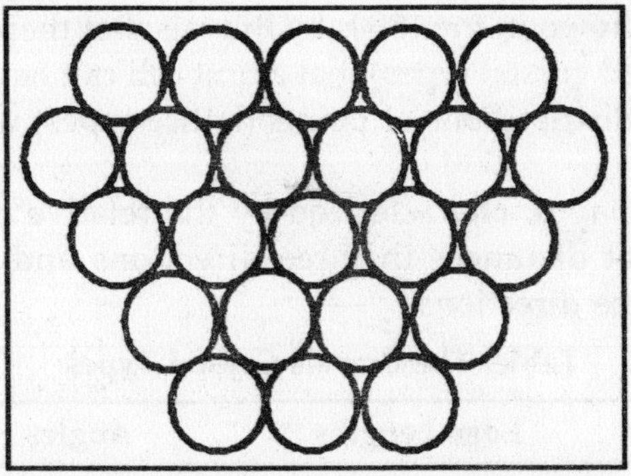

Fig. Layer of Closed Packing

We now add a single sphere of a second layer on top of the first. The added sphere rests in a low point of the first layer, corresponding with a triangular hole, in contact with 3 spheres of the first layer.

The 4 touching spheres lie at the vertices of a regular tetrahedron.

There remains some unfilled space at the centre of this 4-sphere group that is called a tetrahedral hole. Each time we add a sphere to the second layer, we create a tetrahedral hole between the layers.

When additional spheres are added to the second layer, it turns out that they must all go in either "u" holes or "d" holes, but not both. Thus only half of the triangular depressions in the first layer can be used for placement of 2nd-layer spheres.

Triangular holes in the 2nd layer lie directly over the unused holes of the first layer, so that channels pass right through both layers.

The point at the centre of such a channel is surrounded by 6 spheres arranged as 2 equilateral triangles, one triangle in the 1st and one in the 2nd layer.

These 6 spheres form a regular octahedron, a geometrical solid with 6 vertices and 8 equilateral triangular faces. The space surrounded by the 6 spheres is called an octahedral hole. An octahedral hole is larger than a tetrahedral hole because 6 spheres cannot be crowded as closely together as 4 spheres.

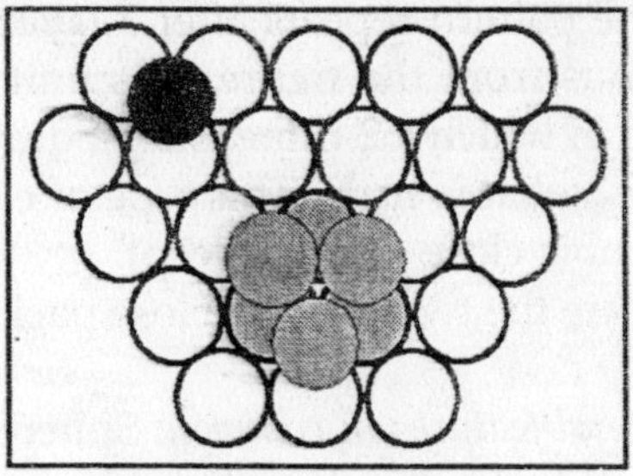

Fig. Black-shaded Sphere

There are 2 ways in which a third layer of spheres can be added to the existing 2 layers in close-packed fashion. In the first way, spheres are place directly over spheres of the first layer. The third and first layers then have spheres in the same locations. The 4th layer then goes over the second, the 5th over the 3rd, and so on, so a pattern that repeats every other layer is obtained. The unit cell is hexagonal in shape; the 1212... close-packing arrangement is therefore called hexagonal close packing (hcp). Note that the coordination number (the number of nearest neighbors) of any sphere in hcp is 12.

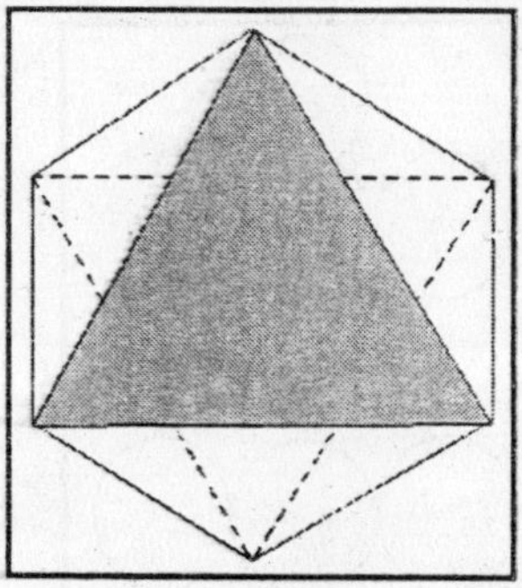

Fig. regular octahedron

The second way of adding a third layer is to place spheres directly over the octahedral holes between the first two layers. In this case, the third-layer spheres do not lie over spheres in layer 1. This is still a close-packing arrangement because the coordination number of any sphere is 12, just as in hcp. The 4th layer of spheres is then placed directly over the spheres in the first layer. This type of close packing is designated 123123...

to indicate that the pattern repeats after 3 layers. Although it may not be obvious from the figure, this unit cell is a cube (this is a situation in which a 2-dimensional drawing does not adequately convey what is true). This type of close packing is therefore called cubic close packing (ccp).

Hcp and ccp are the two ways of close packing spheres in 3 dimensions.

They have several features in common: Spheres occupy 74% of the available space; Each sphere has 12 nearest neighbors that touch it (coordination number);

The number of octahedral holes in the lattice is exactly the same as the number of spheres in the lattice, and there are twice as many tetrahedral holes as spheres! This is not obvious and will be addressed again later.

There are a number of substances that crystallize in one or the other of these two packing arrangements. Beryllium, magnesium, and zinc metal all form hexagonal close packed lattices. Nickel, copper, gold, and aluminum form ccp lattices. Cubic close packing is particularly common. We will take a closer look at cubic lattices now.

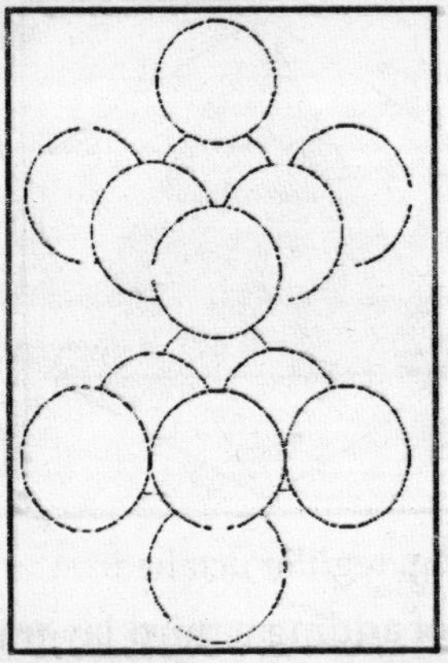

Fig. Unit Cell for 123... Close Packing

The three Bravais cubic lattices are shown in Figure. The simple cubic (sc) lattice is primitive, with points (spheres or atoms) only at the vertices of the cube. Metallic polonium and non-metallic dioxygen (O_2) crystallize with the simple cubic lattice. The coordination number of an atom in the simple cubic lattice is 6.

To determine this, pick one atom in the unit cell; find an atom that appears to be closest to the atom you picked; then count the number of other atoms that are at this same distance from the atom you picked. In counting coordination numbers, you may have to go outside of the single unit cell pictured. In the simple cubic lattice, identical unit cells surround the pictured one on all sides.

Some of the nearest neighbors of your picked atom lie in these neighboring unit cells. Finally, the simple cubic unit cell contains only 1 atom, despite the fact that you see 8 atoms drawn in the figure.

We say that the simple cube has 1 atom per unit cell (1 atom/uc). The procedure for counting atoms in unit cells is described in the paragraph below. The body-centered cubic (bcc) lattice is non-primitive, with points at the corners and the centre of the unit cell (the centre of a cube is called the cube body centre).

Metallic iron and barium crystallize in the bcc lattice. The coordination number is 8 for this lattice (count the nearest neighbors of the body-centre atom), and the body-centered cube has 2 atoms per unit cell. Finally, the face-centered cubic (fcc) lattice is also non-primitive with points at each corner and in the centre of each of the 6 faces of the cube.

Metallic nickel and copper adopt this structure. The coordination number is 12 for this lattice, which means that it must be close packed! The face-centered cubic lattice and the cubic close packed lattice are identical. It is very difficult to see this in two-dimensional drawings. To see it, you must imagine looking along a body diagonal of the face-centered cube.

A body diagonal runs on a straight line from one corner, through the body centre of the cube, to the diagonally opposite corner. As you look along the body diagonal a single sphere; a triangle consisting of 6 spheres; a second triangle of 6 spheres, but rotated 60o with respect to the first one; and finally another single sphere immediately below the first one. The cube is admittedly difficult to visualize from the drawings. The face-centered cubic lattice has 4 atoms per unit cell.

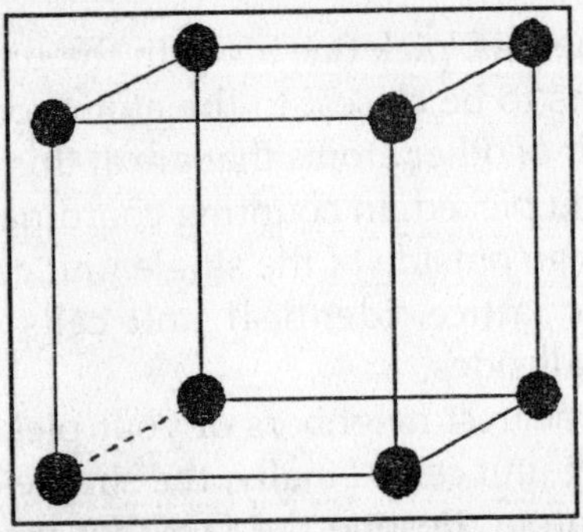

Fig. Three Bravais Cubic Lattices

We now address the problem of determining the number of atoms per unit cell. This is quite simple once it is realized that atoms are generally shared among several unit cells and therefore must be partitioned accordingly.

An atom at a corner (vertex) of a cube is shared among 8 unit cells. To see this, imagine the simple cube to be a building block, and focus on the upper right front corner of the building block. Imagine that this corner is labelled "a". In your mind, place identical blocks on top, to the right, and in front of the block in the figure. Then place additional blocks in the remaining places adjoining these blocks to generate the 8-block arrangement.

Point "a" is now at the centre of this collection, and is simultaneously a corner for each of the 8 blocks; 8 blocks share a vertex. This point cannot be said to belong entirely to any particular block. Instead, each block gets 1/8 of it. The counting rule for a corner atom, then, is that only 1/8 of it is assigned to a single unit cell. Similar thinking leads to the conclusion that a face-centre atom is shared between 2 unit cells, so 1/2 of it is assigned to a particular unit cell; and a body-centre atom is not shared at all, so it belongs entirely to the unit cell that it is found in.

The atom count for a body-centered unit cell then goes like this:

atoms/*uc* = 8 corner atoms × 1/8 + 1 body-centre atom × 1

We will have occasion also to partition points located

along edges of the cube. Edges are the lines connecting adjacent vertices. Since an edge is common to 4 cubes arranged in a square, 1/4 of any point along an edge is assigned to a particular cube. The point-counting rules are summarized in.

Table. Atom Counting Rules for Cubic Unit Cells

Atom location	Fraction Assigned to uc
Corner	1/8
Edge	1/4
Face	1/2
Body-centre	1

If, as we have said, the face-centered cubic lattice is in fact cubic close packed, then there should be tetrahedral and octahedral holes in the fcc lattice. Further, we should find 4 octahedral and 8 tetrahedral holes in a fcc unit cell based on the the last-listed common feature of close-packed lattices above.

To locate these holes it is convenient to place our cube in a coordinate system. The origin of coordinates is located at the lower left front corner of the cube, with the x-axis along the bottom front edge, the y-axis along the left vertical edge, and the z axis along the left bottom edge.

We specify distances along the 3 axes in terms of fractions of the cube edge length. The set of coordinates (1,0,0) means to proceed 1 edge length along x. These are the coordinates of the lower right front vertex of the cube. Similarly, (1/2,1/2,1/2) specifies the body centre of the cube, labelled 2 in the figure. Now examine the face-centered cube. We find the most obvious octahedral hole at the body centre. This point is surrounded by 6 atoms all at the same distance from it. These are the atoms at the face centers of the cube. Similar octahedral holes are located at the cube edge centers. For example, the point labelled 3 is an octahedral hole. It is surrounded at equal distance by the atoms at (1/2,1/2,1), (1,0,1), (1,1,1), and (1,1/2,1/2). The remaining 2 atoms are not part of the unit cell shown, but can still be located with coordinate sets (1 1/2,1/2,1) and (1,1/2,-1 1/2). (Translating these coordinates is good practice in the use of the unit cell coordinate system.) The cube has 12 edges, so there are 12 holes of this type. The counting rules tell us that only 1/4 of each of these edge-centered octahedral holes belongs to the unit cell of interest. The total number of octahedral holes in the unit cell is then 1(body centre) + 12(1/4) = 4. The number of octahedral holes is the same as the number of atoms in the unit cell.

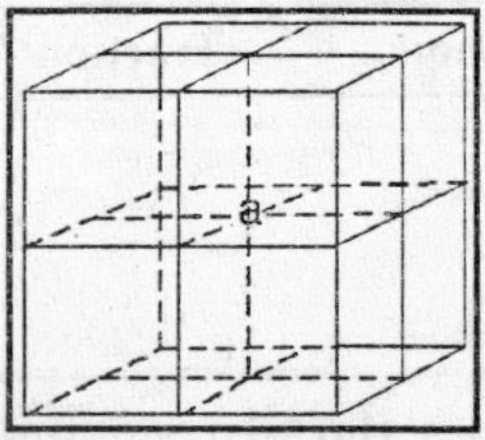

Fig. blocks to generate the 8-block

It is defined by the four shaded atoms, and has coordinates (1/4,1/4,1/4). There are 7 other tetrahedral holes in the unit cell, all with coordinate combinations involving 1/4 and 3/4, and all defined by 1 corner and 3 face-centre atoms. Try to locate them in the figure, and write down their coordinates. Each tetrahedral hole lies entirely within the cube body, so none of them are shared with other unit cells. The number of tetrahedral holes per fcc unit cell is therefore 8, twice the number of atoms in the unit cell. Octahedral and tetrahedral holes are important in the discussion of ionic compounds.

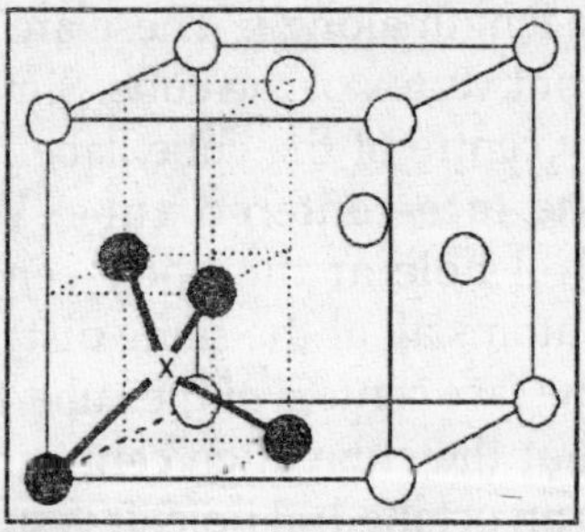

Fig. Tetrahedral Hole in the Fcc Unit Cell

IONIC COMPOUNDS

The ideas discussed thus far apply to crystals in which

the units — spheres or atoms — are all identical. Crystals of pure metals, with close-packed metal atoms, and even crystals of molecular solids (for example, methane, CH_4, in which there is a methane molecule at each lattice point) are examples of real situations in which these ideas apply. Ionic compounds are different because there are two distinctly different units — the cation and the anion — that must be accommodated in the lattice. The resulting lattices are often called compound lattices because they contain two different components.

The cation and the anion differ in charge, of course, and in chemical identity (usually). They also differ in size. Generally speaking, cations are smaller than anions, for easily-understood reasons. Cations contain an excess of protons over electrons. The Coulomb pull of the excess positive charge causes the electron cloud to contract, so that it is smaller than in the corresponding atom (e.g., Na+ is smaller than Na). Anions contain an excess of electrons over protons. The repulsion between the electrons exceeds that in the corresponding neutral atom, causing an expansion of the electron cloud (e.g., Cl^- is larger than Cl). Except for a few rare cases, the cation in an ionic compound is smaller than the anion. In forming a lattice, the space requirements are more important for the larger anion, which tends to dictate which lattice is adopted. The (smaller) cations then fit into the holes, or interstices, of the anion lattice. Many common ionic compounds have lattices that consist of cubic close packed anions, with cations in tetrahedral and/or octahedral holes. Stoichiometries of up to 3 cations per anion can be accommodated in this framework. We now discuss some common ionic lattices.

SODIUM CHLORIDE LATTICE

It can be described as a cubic close packed (face-centered cubic) anion lattice, with Na^+ ions in the octahedral holes. The coordination number (number of nearest neighbors of opposite charge) of each Na^+ ion is 6, and of each Cl^- ion is 6. The cation and anion must have equal coordination numbers because the stoichiometry of the compound is 1:1 (1 cation per anion). The unit cell contains 4 formula units of NaCl (4 Cl^- ions, which define the fcc unit cell; 4 Na^+ ions because there are 4 octahedral holes per unit cell). Many ionic compounds exhibit the NaCl lattice, including most of the group 1 halides, most of the group 2 oxides, and a number of d-metal oxides.

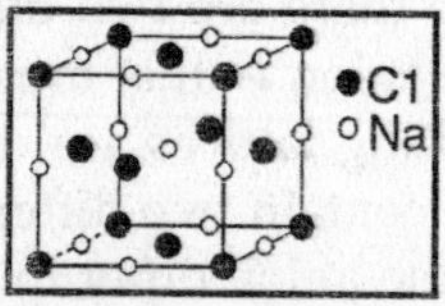

Fig. NaCl lattice

The Zinc Blende Lattice. Zinc blende is the common name for one crystal modification of the common mineral, ZnS. The other form is known as wurtzite. Wurtzite consists of an hcp array of S^{2-} ions with Zn^{2+} ions in one-half of the tetrahedral holes; it will not concern us further here, because we are restricting our attention to cubic lattices. Zinc blende consists of a ccp array of S^{2-}, with Zn^{2+} in half of the tetrahedral holes. Since there are 8 tetrahedral holes in the sulfide ion lattice, but only 4 S^{2-} ions, only half of the holes are used to accommodate 4 Zn^{2+} ions. This is consistent with the 1:1 stoichiometry of the compound. The zinc ions occupy tetrahedral holes so that they can be as far apart from one another as possible to minimize repulsions between positive ions. If the first Zn2+ ion occupies the hole at (1/4,3/4,1/4), the other three will be in holes at (3/4,1/4,1/4), (1/4,1/4,3/4), and (3/4,3/4,3/4). You should locate these holes and verify that the Zn^{2+} ions describe a tetrahedron. The cation and anion coordination numbers are both 4 in zincblende. As for NaCl, there are 4 formula units per unit cell. Other ionic compounds that adopt the zinc blende lattice include BeS, CdS, HgS, and AgI.

CALCIUM FLUORIDE LATTICE

This compound has formula CaF_2, and exhibits the lattice. The Ca^{2+} ion is virtually the same size as the F^- ion, one of those rare situations referred to above, and forms a face-centered cubic lattice. The F^- ions fill all of the tetrahedral holes in the cation lattice. Since there are 8 such holes per 4 Ca^{2+} ions, the stoichiometry is nicely accommodated. The coordination number of the Ca^{2+} ion is 8, and that of the F^- ion is 4. There are again 4 formula units per unit cell. The fluorite structure is very common for ionic compounds of 1:2 (or 2:1) stoichiometry.

The Cryolite Lattice. Cryolite is the common name for the mineral, Na_3AlF_6. Though rare, this substance is of extreme commercial value, because it is used in the process of extraction of aluminum from bauxite ore. Cryolite consists of Na+ cations and AlF_6^{3-} anions. It adopts a lattice in which the anions are cubic close packed (fcc) and the cations occupy all of the octahedral and tetrahedral holes.

The Diamond Lattice. Diamond is one crystal modification of the element carbon. It is, of course, not ionic, nor even a compound. It is dealt with here because it, too, is based on a fcc unit cell. The diamond lattice is just like the zinc blende lattice, with carbon atoms in place of both Zn^{2+} and S^{2-} ions. We therefore describe it as a fcc array of carbon atoms with carbon atoms in one-half the tetrahedral holes. The coordination number of each carbon atom in diamond is.

CESIUM CHLORIDE LATTICE

It is a more open lattice than those considered thus far. This lattice consists of a simple cubic array of chloride ions, with Cs^+ ions in cubic holes. A cubic hole is the space at the body centre of a simple cubic unit cell. There is 1 formula unit of CsCl per unit cell, and the coordination number of both cation and anion is 8. Ionic compounds that adopt the CsCl lattice include CsBr, CsI, the halides of thallium(I), and NH_4Cl.

RADIUS RATIO RULES

Compounds of 1:1 stoichiometry commonly adopt one of three lattices discussed above: the zinc blende lattice, the rock salt lattice, or the cesium chloride lattice. Of interest now is whether or not we can predict, or at least understand, why a particular compound favors one lattice over the others. A number of factors are important in choice of lattice, but one of these dominates in importance: the relative sizes of cation and anion. Because we view the cation as occupying holes in the anion lattice, we begin by calculating the sizes of the various hole types in relation to the size of the anion involved.

The most straightforward calculation is for the octahedral hole.The figure shows one face of the fcc unit cell, with the anions just touching along a face diagonal of the cube. There is an octahedral hole at the centre of each of the 4 cube edges shown; we focus on the right edge. The size of the octahedral hole will be taken as the distance from the edge centre to, say, the top right anion, along the cube edge. (The distances from the hole to the bottom right anion and to the face-centered anion are the same.) We can express the right edge length as a sum of anion and hole radii as in equation:

$$a = 2r- + 2r+$$

Here r- is the anion radius, r+ the hole radius. (The + designation is used in anticipation of putting a cation in the hole.) Similarly, we express the length of the face diagonal (fd) in terms of anion radii.

This is easy, since the anions touch along this line:

$$fd = 4r-$$

Finally, we relate the fd and edge length via the Pythagorean Theorem:

$$(fd)2 = 2a2$$

Take the square root of both sides of 7-4-3:

$$fd = a(2)1/2$$

Substitute 7-4-4 in 7-4-2 and rearrange for a:

$$a = 2(2)1/2\ r-$$

Substitute for a in 7-4-1 and solve for r+:

$$r+ = ((2)1/2 - 1)\ r- = 0.414\ r-$$

The octahedral hole is somewhat less than half as large as the anions that create the hole. By similar geometric methods, equations , for tetrahedral and cubic holes respectively, can be developed:

$$r+ = 0.224\ r-$$

$$r+ = 0.732\ r-$$

You should try to obtain these expressions, making use

of right triangles and Pythagorean relationships. (In addition to equation which relates the face diagonal and edge length of a cube, equation relates the cube body diagonal (bd) to the edge length:

$$bd = (3)1/2a$$

According to Equations , a tetrahedral hole is about 1/2 the size of an octahedral hole and about 1/3 the size of a cubic hole. Only cations that are relatively small compared with the anion can be accommodated in tetrahedral holes. As cation size increases relative to anion size, the cation will prefer a larger hole, with a larger coordination number of anions. Equations are the basis for the following radius-ratio rules.

If the ratio of the cation and anion radii (r+/r-) lies in the range between 0.224 and 0.414, the cation prefers to occupy a tetrahedral hole.

If $r+/r-$ lies in the range between 0.414 and 0.732, the cation prefers an octahedral hole.

If $r+/r-$ exceeds 0.732, the cation prefers a cubic hole.

These rules make good sense. If $r+/r- < 0.224$, the cation is too small for the tetrahedral hole. The anions will be touching, but will not be touching cations. Since it is the cation-anion attractions that stabilize the lattice, whereas anion-anion or cation-cation repulsions tend to destabilize it, this is not a favorable situation. When $r+/r- = 0.224$, the cation just fits the tetrahedral hole. Although anions are still touching, they now touch cations as well.

This is a stable arrangement. As $r+/r-$ becomes > 0.224, the cation is too big for the tetrahedral hole and starts to push the anions apart. The cation-anion attractions are maintained, because cations and anions still touch, but now anion-anion repulsions are decreased because the anions no longer touch. This is an even better arrangement than the perfect fit. This situation continues to improve until $r+/r-$ becomes equal to 0.414. At this value, the cation prefers to occupy an octahedral hole because it can increase its coordination number without increasing anion-anion repulsion to any significant extent.

The number of cation-anion attractive interactions increases from 4 to 6, giving a more stable lattice. As r+/r-increases between 0.414 and 0.732, cations in octahedral holes maintain their contact with 6 anions while pushing the anions further and further apart, decreasing repulsions between them. Throughout this range of *r+/r–*, occupation of an octahedral hole becomes increasingly favorable until the radius ratio becomes 0.732.

The cation can now increase its coordination number to 8 anions, again increasing the number of attractive cation-anion interactions.

Table. Radius Ratio Rules

Range of r+/r-	Lattice Type	Example
0.224-0.414	ZnS	ZnS
0.414-0.732	NaCl	KCl
> 0.732	CsCl	CsCl

Ion radii and the radius ratio rules can be used to predict the probable choice of lattice for an ionic compound.

The radius ratio rules provide useful guidelines to the lattice structures of ionic compounds. It is important to realise, however, that they are not hard and fast; the predictions made using the rules are often wrong. For example, in the zincblende form of ZnS, Zn^{2+} ions occupy tetrahedral holes in a fcc lattice of S^{2-} ions. In fact, this compound is the prototype for this type of structure.

However, the radius ratio rules predict that the Zn^{2+} ions should occupy octahedral holes! The radius ratio rules fail frequently because the relative size of the ions is not the only factor that is important in determining structure. Of particular importance is the tendency toward at least some covalent bonding between cation and anion. Covalency involves directional requirements and may influence the type of hole preferred by the cation.

X-RAY CRYSTALLOGRAPHY

In 1912, von Laue predicted that crystals could serve as diffraction gratings if the wavelength of radiation used were approximately the same as the distance between planes of atoms in the crystals.

This spacing was known to be on the order of 1×10^{-10} m, corresponding to wavelengths in the x-ray region of the electromagnetic spectrum. Subsequently, WH Bragg discovered that when X-rays of appropriate wavelength are directed onto the smooth face of a crystal, the X-rays are diffracted and a pattern characteristic of the lattice structure of the crystal is obtained.

This observation is the foundation for the modern technique of X-ray Crystallography, the most powerful method that we have for determining molecular structure. X-ray crystallography is based on the interaction of light with matter. Unlike spectroscopy, however, the light is not absorbed by the matter in this case.

Instead, it is scattered from the atoms of the crystal in a very definite way. Analysis of the scattering enables the arrangements of atoms within the crystal to be deduced. The analysis of x-ray diffraction patterns is complex, even for materials of simple structure.

However, thanks to the efforts of many crystallographers over many years, it is today done quite quickly and routinely. Once a suitable crystal of a substance is grown, the modern crystallographer is able to mount it in a fully automated x-ray diffractometer, the instrument that is used for the x-ray diffraction study.

The diffractometer automatically obtains all of the necessary data for the determination of the lattice structure. The data are then fed to one of a number of computer programs that have been developed and refined for the analysis of such data, and the structure is "solved." The output of the programme reveals not only the lattice structure, but also the locations of all of the atoms of a molecule of the substance. Structures of substances ranging from simple salts or molecular substances to huge complex proteins and enzymes can be determined using this approach. The structures of two molecules, obtained from x-ray diffraction data.

THE BRAGG EQUATION

A source of X-rays at the left emits radiation, two beams

of which are shown impinging on the planes of atoms and scattering off to the right, where they are "seen" at a detector. The angles made by the incident beam and the deflected beam with the planes of atoms are the same.

A diffraction pattern is a series of alternating light and dark lines (sometimes spots). The white spots result from the arrival at the detector of reflected beams that are all in-phase. That is, their waves are matched up peak for peak and trough for trough.

Waves that are in phase add together to give a wave with larger peaks and larger troughs — i.e., more intensity. This is called constructive interference. The white spots in the diffraction pattern correspond to deflection angles that give in-phase waves.

Similarly, the dark areas correspond to angles that give out-of-phase waves. Here the peak of one wave arrives at the detector at the same time as a trough of a second wave arrives. Such waves cancel each other, and very little or no signal is seen at the detector. This is called destructive interference.

The criterion that must be met in order that two waves arrive at the detector in phase is that the difference in the distances travelled by them must correspond to an integer number of wavelengths. This will give the maximum intensity at the detector.

If the difference in distance travelled corresponds to a half-integral number of wavelengths, cancellation will occur, and there will be no intensity registered at the detector. The difference in the distances travelled by the two deflected beams is 2x. Since x is related to the angle of incidence (deflection) and to the separation of the planes of scattering atoms as x = dsin?, and since 2x must be an integer number, n, of wavelengths in order to maximize intensity, equation 7-5-1 results.

$$n = 2d\sin$$

This is known as the Bragg equation. It is the basis of modern x-ray crystallography. It shows that if we measure the angles at which constructive interference occurs, we can calculate d, the spacing between atom planes in the crystal.

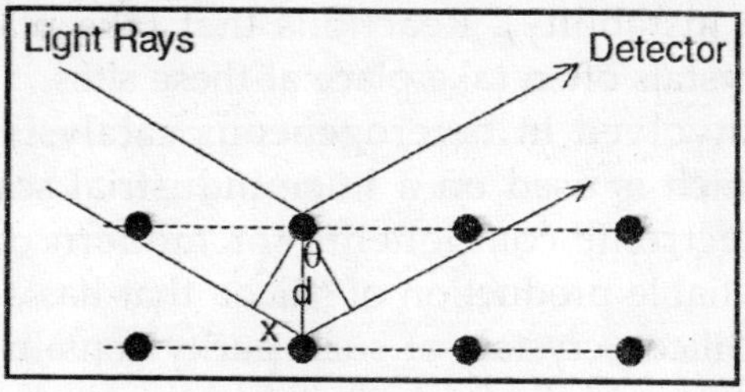

Fig. Atoms in a Crystal

Lattice Imperfections

Perfect crystals, in which every atom or molecule is in place, are an idealization, like the ideal gas. Real crystals contain fairly large numbers of imperfections, called defects. Three major types of defect are called point defects, line defects, and plane defects. Common point defects in simple lattices (in which all lattice points are identical) are vacancies (in which an atom is missing); atoms out of place, occupying what is normally a hole in the lattice; substitutional impurities, in which foreign atoms occupy occasional sites; and interstitial impurities, in which foreign atoms occupy holes in the lattice. Ionic (compound) lattices exhibit Schottky defects, in which a cation or anion is missing; and Frenkel defects, in which an ion leaves its site and enters a lattice hole.

Line defects result from missing rows of atoms. Particularly common are edge defects, in which an extra partial plane of atoms is inserted near an edge of the crystal. Plane defects in crystals occur at the interface between two differently-oriented crystalline regions. As might be expected, such defects greatly diminish the resistance of the crystal to fracture.

Crystal defects are important for a number of reasons. A problem with defects, especially of the line and plane type, is that they diminish the structural strength of the solid. Annealing is used to offset this. Rusting of iron often occurs at defect sites on the surface, which may result when the metal is worked or penetrated by a rivet or bolt. However, defects can be advantageous as well. Defects are sites of high energy (i.e., relative instability). Reactions that take place at the surfaces of crystals often take place at these sites.

This is involved in heterogeneous catalysis by solid materials, which is used on a huge industrial scale in our economy. Electronic components for modern computers require the reliable production of silicon that has a purity of 99.999999%. Silicon crystals of such purity contain very few defects, which would diminish their functional effectiveness. Silicon of this purity is produced by a process called zone refining. In this method, a bar of silicon of high purity is passed very slowly past a very narrow heating element, which melts a narrow region of the silicon bar. Impurities in the silicon dissolve in the melted part, and follow along with the melted region as the bar moves past the heating element. Eventually the fused region, containing all of the impurities, reaches the end of the bar, where it is solidified and cut off. The remaining silicon has the purity required in the manufacture of electronic components.

Properties of Liquids

We take up the liquid phase last because it is the most complex and least understood of the phases of matter. Liquids have properties intermediate between those of the solid and gas phases: A sample of liquid has a fixed volume (like a solid) and a molar volume only slightly greater than that of the solid. However, A liquid conforms to the shape of its container, like a gas.

A liquid flows readily, like a gas: Liquids are relatively incompressible. The relative compressibilities of the three phases are in the order g >> l > s. The incompressibility of liquids finds practical use in hydraulic machinery, such as automobile brakes.

Diffusion is slower in liquids than in gases, but faster than in solids. The relative rates of diffusion in the three phases are in the order g >> l > s.

These properties are readily observed visually. Several other properties of liquids are more subtle:

Liquids give x-ray diffraction patterns that are diffuse,

rather than sharp like those of solids. The pattern implies some order in the molecular arrangement, but less than in the solid.

- Liquids exhibit surface tension.
- Liquids exhibit viscosity.

Liquids in closed containers evaporate to an extent. The resulting vapour exerts a pressure called the vapour pressure, Pvap, of the liquid. The magnitude of Pvap for a particular liquid depends only on temperature, not on the container size or the amount of liquid present. Liquid and vapour exist in equilibrium.

MODEL OF LIQUID PHASE

The small molar volume and incompressibility of liquids suggest that the molecules are close together. However, flow and the lack of a fixed shape indicate molecular mobility, despite the closeness. A feature of the model that is not evident from the figure is that the molecules are in constant motion, as in the gas and solid, colliding frequently with neighbors. As for the gas and the solid phases, the kinetic energies of the particles of the liquid are described by the Maxwell-Boltzmann distribution, and the average KE per particle is 3/2 kT. Each molecule is fairly closely surrounded by other molecules, but there are definite gaps (holes) in the structure. Molecules use these gaps to slip and slide easily past one another, which manifests macroscopically in the ability to flow.

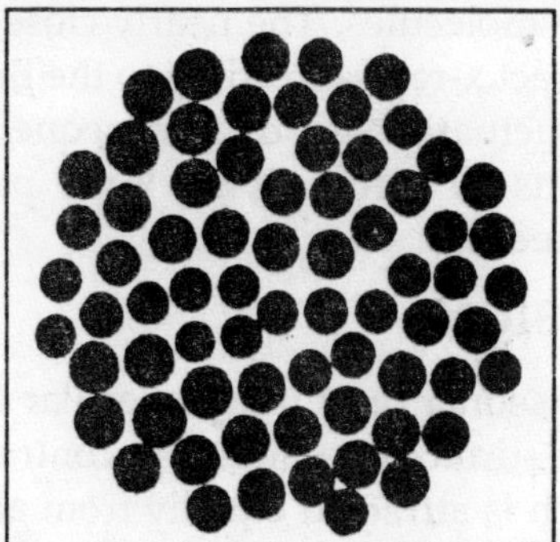

Fig. Model of the Liquid Phase

There are regions in the liquid that are quite ordered,

similar to the solid, but the regions are constantly shifting position as molecules move to close gaps and open new ones. The gaps prevent the order from extending over long distances.

Thus liquids have short-range order, in contrast to the long-range order of crystalline solids. The gaps allow diffusion of a molecule through the liquid, but frequent collisions with neighboring molecules makes diffusion slow. Contrast this with the rapid diffusion in the gas phase, where a molecule travels a long distance between collisions; and the extremely slow diffusion of the solid phase, where a molecule is locked into its lattice location.

Diffusion rates in solid, liquid, and gas are best understood in terms of the mean free path, the average distance travelled by a molecule between collisions. The mean free path in the solid is virtually zero. In the liquid, it is a fraction of the molecular diameter. But in the gas, the mean free path may be 10-100 molecular diameters, depending on gas pressure.

INTERESTING PHYSICAL PROPERTIES

DIFFUSE X-RAY DIFFRACTION PATTERNS

Within these regions, molecules are almost close-packed. However, the gaps allow fluctuation in the distance of separation of the molecules. The nearly close-packed regions of the liquid diffract x-rays according to the Bragg equation, n = 2dsintheta . Fluctuations in d, a consequence of the gaps, cause fluctuations in theta, giving x-ray patterns that are diffuse, or smeared out.

SURFACE TENSION

This is a consequence of intermolecular forces. Consider a molecule at the surface of a liquid. In contrast to a molecule in the bulk, which is attracted equally from all directions, the surface molecule suffers an imbalance of forces because there are no molecules above it.

The net force on it is directed toward the interior of the

liquid. This pulling in of the liquid surface is called surface tension, symbolized z. The surface molecule has higher potential energy than a bulk molecule because of the lesser force on it.

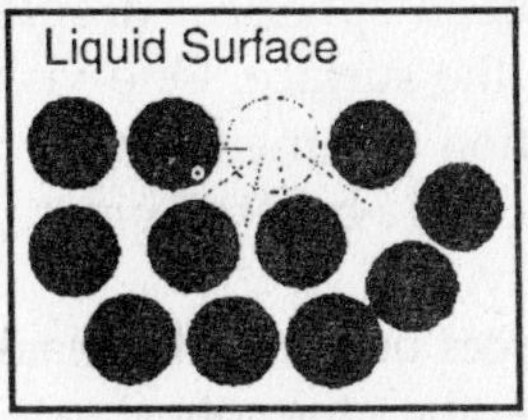

Fig. Surface Tension of liquids

To minimize energy, the liquid adopts a shape that minimizes the number of surface molecules. The shape with the smallest surface area-to-volume ratio is the sphere. Thus surface tension explains the spherical shape of raindrops and water droplets on a freshly waxed car: the spherical shape minimizes potential energy.

It is a safe prediction that manufacture of ball bearings will someday be carried out in space, where gravitational forces will not distort the sphericity that surface tension imposes on a drop of molten metal. As temperature increases, molecular kinetic energy can overcome the attraction of intermolecular forces, and surface tension decreases according to equation.

$$= a - bT$$

The constants a and b are characteristic of a particular liquid. For ethanol, a = 24.05×10^{-3} J/m^2 and b = 0.0832×10^{-3} J/ m2-oC. The units of surface tension are energy per area, consistent with the relation between potential energy and surface area discussed above.

A drop of water on a clean glass surface spreads and flattens, forming a film on the surface. This phenomenon of spreading and flattening is called wetting. Wetting occurs because forces between the surface molecules and molecules of the liquid exceed the intermolecular forces within the liquid. The liquid spreads to maximize the area of contact with the surface.

Forces between unlike molecules are called adhesive forces; those between like molecules are called cohesive. Wetting occurs because adhesive forces exceed cohesive forces. If the reverse is true, a liquid will not wet a surface, but instead will sit on the surface in spherical droplets to minimize the area of contact with the surface. One visible consequence of wetting is the formation of a meniscus at the top of a column of liquid in a tube, such as a buret or pipet. The menisci formed by water and mercury.

The adhesive forces between the glass surface and water molecules are stronger than the cohesive forces between water molecules. Consequently, water achieves a maximum surface area of contact with the glass by bending up at the edges and sagging in the centre.

The reverse is true for mercury, which minimizes the surface area of contact with the glass and forms a spherical surface shape to maximize cohesive forces.

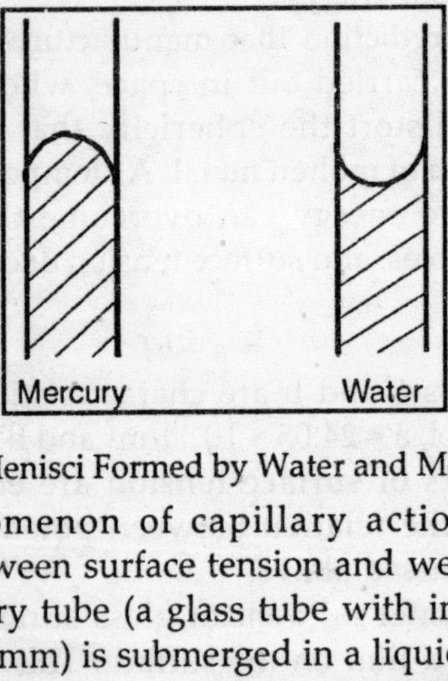

Fig. Menisci Formed by Water and Mercury

The phenomenon of capillary action results from competition between surface tension and wetting. When one end of a capillary tube (a glass tube with inner diameter of approximately 1mm) is submerged in a liquid, a difference in height develops between the levels of liquid in the capillary and in the bulk liquid.

A liquid that wets glass rises in the tube; a liquid that does not wet glass falls. As water wets the inside wall of the capillary, the surface area of the water is increased.

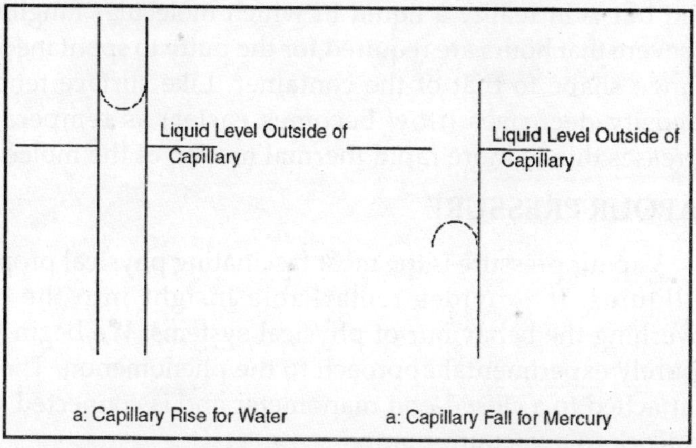

Fig. Capillary Action

However, surface tension tends to minimize the number of molecules at the surface, and causes the water to move up the tube, against the force of gravity, to offset the surface area increase resulting from wetting. For mercury, adhesive forces are weak, so the level in the capillary falls to minimize contact with the glass surface, and the surface of mercury in the capillary is pulled into a sphere by the strong cohesive forces. The result is capillary fall. Capillary action is important in the operation of blotters and sponges and is the mechanism by which water penetrates soils.

VISCOSITY

The viscosity of a liquid is a measure of its resistance to flow. A mobile liquid is one that flows readily (e.g., water). A liquid that flows with difficulty (e.g., molasses) is said to be viscous. Liquids composed of small, roughly spherical molecules are mobile, because the molecules can roll over one another smoothly, like ball bearings. Viscous liquids are composed of long, string-like molecules that easily become tangled during flow.

For example, the molecules of high viscosity motor oils consist of long chains of carbon atoms that intertwine and impede flow of the liquid. An extreme example of molecular tangling occurs in silly putty, a polymer that appears to be a solid but is in reality a liquid in which molecular tangling is so severe that hours are required for the putty to spontaneously change shape to that of the container. Like surface tension, viscosity decreases (flow becomes easier) as temperature increases due to more rapid thermal motion of the molecules.

VAPOUR PRESSURE

Vapour pressure is the most fascinating physical property of liquids. It provides remarkable insight into the laws governing the behaviour of physical systems. We begin with a purely experimental approach to the phenomenon. The box is attached to a closed-end manometer and is connected via a stopcock to a vacuum pump.

A thermometer is inserted in the box and there is a septum-capped port (opening) that can be penetrated with a syringe needle. The stopcock to the vacuum pump is opened and the air is pumped out until the two arms of the manometer are at the same level. The stopcock is then closed. This is the initial situation, state 1, shown at the top left of the figure. Several experiments will now be performed in sequence.

Expt 1. The box is thermostatted at 25°C, and 5.000 g (approximately 5 mL) of water is injected through the septum by syringe. We observe that a pressure develops within the box. Over a brief time (perhaps 1 minute) the pressure rises to 23.8 torr and remains there. Note that the rate of change of the pressure (the slope of the curve) is largest just after injection of the liquid, and gradually decreases as the pressure smoothly approaches the final value. There is still liquid at the bottom of the box. We conclude that some water has evaporated to form water vapour, which is responsible for the observed pressure. Since the pressure is caused by the vapour formed by evaporation of liquid, it is called the vapour pressure, Pvap. Assuming that water vapour behaves as an ideal gas, we can use the ideal gas law to calculate the moles of water in the vapour phase. The result is 6.40×10^{-4} moles, or 0.0115 g, of water in the vapour phase.

Expt 2. With T = 25°C and the vapour pressure steady at 23.8 torr, we inject 5.000 g more liquid water to the box. This time we observe no change in the pressure registered at the manometer. This is the situation at state 3 of the diagram. We conclude that the value of Pvap does not depend on how much liquid water is in the box.

Expt 3. It is identical in all respects to the first box except that its volume is 1.00 L. The 1.00-L box has been evacuated by pumping until the Hg level in both arms of the manometer is the same. With the box at 25°C, we inject 5.000 g H_2O. As for Expt 1, the pressure rises over several minutes to a final value of 23.8 torr, according to a curve. The moles of water in the vapour phase must be twice as great as for the original box, because the volume is twice as great. But the pressure is the same. We conclude that Pvap is independent of the container size.

Expt 4. The box contains 6.40×10^{-4} moles of water vapour, exerting a vapour pressure of 23.8 torr, and a little less than 10.00 g liquid water, all at 25°C. We now quickly raise the temperature of the box to 35°C. Over a period of a few minutes, the pressure rises to 42.18 torr. The change of pressure with time. This is the situation in state 4. By drawing off the liquid and weighing it, we find that 9.980 g of liquid remains. We conclude that raising the temperature has caused more liquid to evaporate, giving a larger Pvap.

Expt 5. Beginning with state 4, we raise T to 45°C. The pressure rises again, levelling off at 71.88 torr. Change of pressure with time follows a curve.

Expt 6. Beginning with state 5, we lower T to 25°C. Over several minutes, pressure decreases to 23.8 torr, following a time curve. Simultaneously, some water vapour condenses to liquid, as we ascertain by weighing the liquid. We are now back to state 3. Experiments 4-6 tell us that Pvap increases with temperature. Moreover, the increase is non-linear: the increase between 35 and 45° (29.7 torr) is much greater than that between 25 and 35o (18.38 torr). We will do one last experiment using water.

Expt 7. Remove all liquid water from the 0.500-L box, and pump out all water vapour. The box is empty with equal levels of Hg in the manometer arms. This is state 6. We inject 0.00700 g H_2O into the box. This time, the pressure increases to a final value of 14.4 torr and all of the liquid disappears. We conclude that after injection, evaporation occurs as the water attempts to achieve a pressure of 23.8 torr, the characteristic Pvap at 25°C. However, before this pressure can be reached, all of the water has evaporated. The amount of liquid injected was insufficient to achieve the characteristic Pvap. The minimum quantity of water needed to achieve P = 23.8 torr was found in Expt 1 to be 0.0115 g. Any amount less than this will evaporate entirely, giving a final pressure equal to the value calculated using the injected number of moles in the ideal gas law.

Expt 8. Empty the 0.500-L box and inject 5.000 g of a different liquid, say ethanol (ethyl alcohol, C_2H_6O) at 25°C. We observe the same pattern of behaviour as with water, except that the final vapour pressure is 54.24 torr. We conclude that all liquids have a tendency to evaporate to some extent in a closed container, and establish a vapour pressure that depends only on T and the identity of the liquid. The vapour pressure phenomenon is a universal property of liquids.

Before summarizing the results of our thought experiments, we deal with a related matter. Suppose we put 10 mL of water in a beaker and leave it open to the atmosphere. Over a few days, the water evaporates completely. (Acetone and ethyl alcohol would completely evaporate in a few hours; diethyl ether in a few minutes.) Why? This situation differs from that involving the water in the box because the beaker system is not closed — that is, there is no fixed boundary on top of the beaker to contain the water vapour. Water in the beaker evaporates in an attempt to establish the characteristic Pvap (23.8 torr at 25°C). Water vapour develops over the surface of liquid in the beaker, but begins to diffuse away because there is no confining boundary on top. This diffusion, aided by air currents in the room, moves vapour molecules away from the space over the beaker. This reduces the local water vapour pressure, and more water evaporates to make up for the loss. This process continues until all liquid is gone.

The results of our experiments with vapour pressure can be summarized as follows: A liquid in a closed container evaporates to some extent. The resulting vapour exerts a characteristic vapour pressure, Pvap. The value of Pvap is independent of the container size and the amount of liquid, provided that the amount exceeds a critical minimum. Pvap increases steeply with T: Pvap is a function only of the chemical identity of the liquid and the temperature. Liquids have a natural tendency to evaporate to some extent to form vapour.

The thought experiments, which can be verified readily in the laboratory, provide much descriptive information about the vapour pressure phenomenon. But the questions of why liquids tend to evaporate, and why they establish characteristic vapour pressures, are unanswered by the experiments.

DISSOLVING POWER OF LIQUIDS

Liquids are invaluable to chemists as solvents. They are able to accommodate molecules of solute readily within the gaps in their structures to produce solutions that are themselves liquids. Such solutions are readily poured or otherwise transferred from one container to another; are conveniently measured out quantitatively using volumetric apparatus; and provide a uniform medium in which a chemical reaction can occur. The unique ability of liquids to act as solvents is due to their dissolving power — the molecular motion of the liquid dislodges ions or molecules from the surface of crystals of a solid, then accommodates the separated ions or molecules in gaps. In this section we briefly discuss the dynamics of the dissolution process at the molecular level.

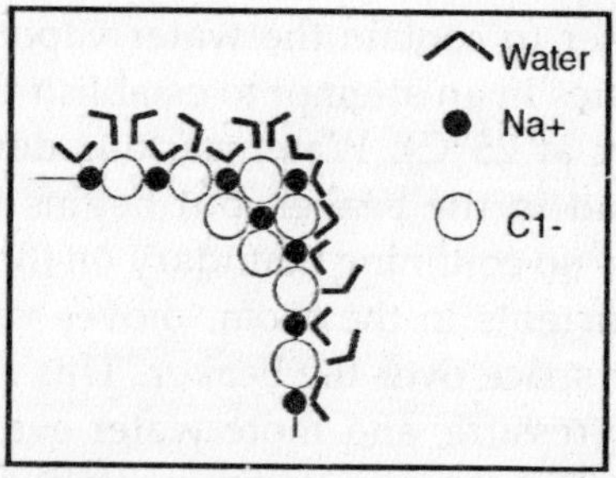

Fig. Microscopic View Adjoining Faces of a Crystal of NaCl

Na^+ ion on a face of the crystal differs from those in the interior because it experiences forces of attraction from only 5 nearest-neighbour Cl^- ions, rather than 6. Facial Cl^- ions are in a similar situation. Such ions are therefore in shallower potential wells than ions in the interior of the crystal and should be more easily dislodged. Edge ions, such as the Na^+ ion labelled "a" in the figure, are attracted by only 4 nearest neighbors, and should be even more readily dislodged than face ions. Water molecules from the solvent continually bump against the faces and edges of the solid crystal.

As they approach the crystal, their dipoles become oriented due to interactions with the face ion that they approach. Water molecules approaching cations have the O atom oriented toward the crystal face; those approaching anions put a H atom forward. When a rapidly moving water molecule strikes a face or edge ion, sufficient kinetic energy can be transferred to dislodge the ion, which moves away from the crystal into a gap in the solvent structure (e.g., Na^+ ion "b" in the figure).

It becomes immediately surrounded by a cage of water molecules, with oxygen atoms nearest the Na+ ion to maximize ion-dipole forces. The solvent cage insulates it from the crystal and makes return to the crystal face unlikely. Since edge ions are more readily dislodged than facial ions, and because they can be approached by solvent from several directions, dissolution occurs more rapidly at crystal edges than faces. The initially sharp edges of a dissolving crystal become rounded as dissolution proceeds.

PHASE TRANSFORMATIONS

A phase transformation is a conversion of a pure substance from one phase to another. There are 6 types, each involving two of the three phases:

- *Melting (fusion):* Conversion of solid to liquid.
- *Freezing:* Conversion of liquid to solid; the reverse of melting.
- *Vaporization:* Conversion of liquid to vapour (gas). Its reverse is
- *Condensation:* Conversion of vapour to liquid.
- *Sublimation:* Conversion of solid to vapour. Its reverse is
- *Deposition:* Conversion of vapour to solid.

Since the solid and liquid phases have small molar volumes due to close packing of molecules, they are referred to as condensed phases.

The two variables that determine the phase in which a substance exists are the temperature and the pressure to which the substance is subjected.

Phase transformations may therefore be accomplished in two limiting ways: by changing the temperature of the substance while keeping it under constant pressure; or by changing the pressure on the substance while keeping it at constant temperature. The first approach is the more familiar of the two. We discuss that now and postpone a discussion of pressure-induced phase changes.

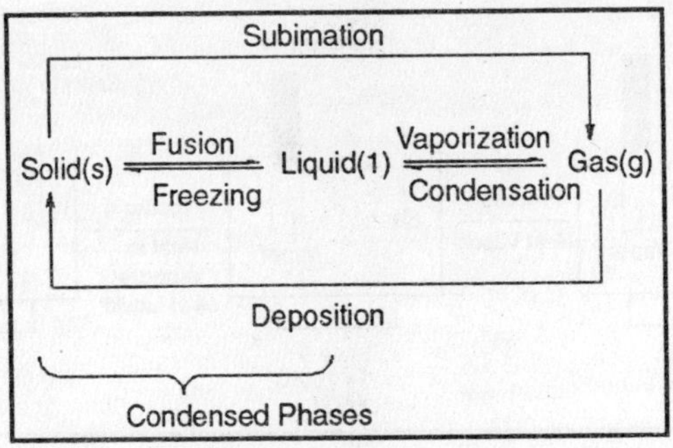

Fig. Phase Changes

We make three assertions about temperature-induced phase changes: At a constant pressure, a phase transformation occurs at a constant temperature that is characteristic of the substance.

To convert a given amount of substance from solid to liquid to gas requires addition of energy. This is usually accomplished by adding heat. The energies of the phases are therefore in the following order:

$$Es < El < Eg$$

Conversion from solid to liquid to gas decreases the order (degree of organization) of the substance.

To clarify these assertions, we take a thought experimental approach to the phase changes of a substance, much as we did for vapour pressure.

The results of these thought experiments can be readily verified in the laboratory. We start with a quantity of solid water (ice) in a syringe. The initial temperature of the ice and the inside of the syringe is -10°C. We plan to slowly add energy to the ice by controlled heating, which can be done by submerging the syringe in a large antifreeze bath with a hot plate under it.

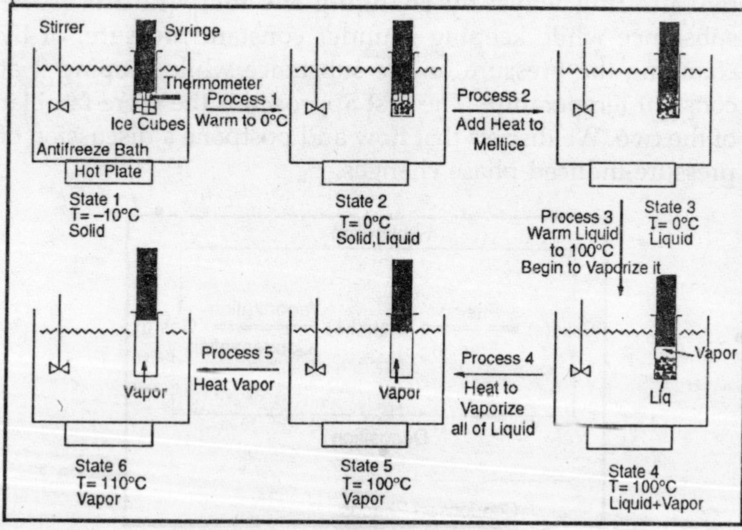

Fig. Phase Transformations

As we slowly add heat, we monitor the temperature of the water in the syringe with a thermometer inserted through an air-tight seal in the syringe. We begin at state 1.

Process 1. Turn on the hot plate and the bath stirrer. As the ice warms slowly from -10 to 0°C, nothing seems to happen. Temperature changes slowly and steadily. When the ice reaches 0°C, it begins to melt, and we see some liquid water form in the syringe.

Process 2. We continue to gently heat the bath. The ice gradually melts, but the temperature of the ice-water mixture remains constant at 0°C, despite the addition of heat. The temperature of the bath rises above 0°C, but as long as both ice and liquid water are in the syringe, T in the syringe remains at 0°. State 3 is reached when the last ice crystal melts. All water in the syringe is liquid, at T = 0°C.

Process 3. Once the ice has melted, the temperature of the water in the syringe begins to rise. As we slowly warm the bath, the water temperature slowly rises. Nothing happens in the syringe other than a slight increase in the volume of the liquid until T = 100°C. Then water begins to vaporize. This is the situation at state 4.

Process 4. We continue to slowly add heat to the bath and observe that temperature in the syringe remains constant at 100°C as long as both liquid and vapour are present, even though the bath temperature rises above 100°C. We also observe that as vapour forms, it pushes the plunger out. After all liquid has vaporized, at state 5, the temperature of the water in the syringe once more starts to rise.

Process 5. Continue heating the bath to T = 110°C. The temperature of the vapour in the syringe rises from 100° at state 5 to 110° at state 6, while its volume increases according to the ideal gas law. The pressure of the water vapour is maintained constant at 1 atm by movement of the syringe plunger.

These experiments verify the assertions at the beginning of the section. First, phase changes occur at constant T. This is shown by processes 2 and 4. The fixed, characteristic temperature at which a substance melts is called the melting point (s to l) or freezing point (l to s), and is symbolized Tf. The normal melting point is the temperature of melting under a pressure of 1.00 atm. For water, this is 0°C. The characteristic temperature at which a substance boils under 1.0 atm pressure is called the normal boiling point. For water, this is 100°C. Second, process 2 shows that an input of energy is required to convert a solid to liquid. The same is true for conversion of liquid to gas (process 4). For each of these processes, the change in energy, ΔE (Efinal - Einitial) is positive. Processes 2 and 4 may be represented as follows:

$H_2O(s)$, 0°C → $H_2O(l)$, 0°C [E > 0]

$H_2O(l)$, 100°C → $H_2O(g)$, 100°C [E > 0]

The process in equation occurs at constant T. Therefore the change in kinetic energy of the molecules is 0 (recall that an average molecule has KE of 3kT/2 at temperature T, regardless of phase). ΔE is therefore a change in potential energy, ΔPE. Thus the PE of liquid is higher than that of solid water at 0°C. Similarly, the PE of vapour is higher than that of liquid water at 100oC. Combining these conclusions gives equation

PEg > PEl > PEs at any T

PE > 0 for sublimation, fusion, and vaporization; PE < 0 for the reverse processes. Third, the amount of disorder at the molecular level increases in our sequence of processes (for reasons that will become clear later, we phrase this in terms of disorder rather than order). With addition of thermal energy (heat), the molecules convert from the long-range organized arrangement of the solid to the completely chaotic and random molecular distribution of the gas. Even warming the gas from 100 to 110oC decreases the order, because the volume available for the molecules to bounce around in increases.

disorder s < disorder l < disorder g

Phases and the Potential Well, Revisited

The potential well in terms of electromagnetic forces within or between the fundamental units of substances. We concluded that forces are strongest and potential energy is minimized when fundamental particles form an orderly close-packed arrangement, and we recognized this as the situation in the solid phase. Heating the solid phase increases the kinetic energies of the particles of the solid phase until they are able to partially overcome the close-packed forces and separate somewhat. At this point, solid melts to liquid.

Separation has increased the potential energy of the particles. Further heating eventually gives the particles sufficient kinetic energy that they may escape the forces in the liquid into the space above the liquid; i.e., vaporization occurs. In the vapour phase, molecules are so widely spaced that forces between them are negligible and potential energy has increased to zero. We thus see that phase conversion from solid to gas can be viewed in terms of particles climbing out of the potential well. Similarly, in the reverse conversion, particles fall into the potential well. The decrease in potential energy of the molecules is released as heat.

PHASE EQUILIBRIUM

MAJOR CONCEPT AREA

Equilibrium. The concept of dynamic equilibrium is an extremely important one in chemistry. We then apply energy and entropy ideas to the physical equilibria established among the phases of one or more pure substances under varying conditions of temperature and pressure. The concept of dynamic equilibrium is then extended to a variety of situations involving a balance of opposing rates across a boundary between two phases. In all cases we will find that the rate of passage from one phase to another depends on the surface area of contact between the two phases, and the concentration of substance in the phase of origin. This idea will help us to understand equilibria involving 2 (or 3) phases of a single pure substance.

Such equilibria occur all around us in nature and are important in weather and the deterioration of infrastructure. We will then apply the idea to equilibria in which a substance distributes itself between 2 other substances that are in contact with each other; and to equilibria involving the dissolution of molecular and ionic solids and gases in liquids. Solubility of solids and gases is extremely important. For example, fluoridation of water is done to convert the principal structural component of teeth, apatite ($Ca_5(OH)(PO_4)_3$), to fluoroapatite ($Ca_5(F)(PO_4)_3$), which is both less soluble and more resistant to chemical attack by food components (particularly acids!) than is apatite. The dissolution of gases such as CO_2 and O_2 is essential to the aquatic life cycle.

SPECIFIC CONCEPTS

Disorder increases in spontaneous processes: Dynamic equilibrium is a balance of opposing processes occuring at equal rates.

In a closed container, a dynamic equilibrium involving two phases of a pure substance can be established at certain pressures and temperatures; the rate of passage from phase 1 to phase 2 equals the rate of passage from phase 2 to phase 1.

When salts dissolve, a dynamic equilibrium is established between undissolved solid and ions in solution; the rate of dissolution equals the rate of crystallization.

When gases dissolve, a dynamic equilibrium is established between gas above the solution and dissolved gas; the rate of dissolution equals the rate of escape.

QUALITATIVE ASPECTS OF ENTROPY

We have also seen that a major driving force for physical and chemical systems is the tendency toward minimum potential energy. This leaves us with a question: why do solids melt to form liquids, and why do these vaporize to gases? If the solid is the lowest-energy phase, and if all systems seek lowest energy, it seems that all substances should be solids at all temperatures and pressures. But this is clearly wrong. To understand the way pure substances behave requires an additional concept.

An ideal gas is confined in bulb A, connected via a closed stopcock to evacuated bulb B. What will happen when the stopcock is opened? Experience tells us that the gas will expand to occupy bulb B until the pressure in both bulbs is uniform. Why does this happen? An initial response might be that by expanding, the gas decreases in potential energy. But closer examination reveals that this is not true. First, if the gas expands at constant temperature (which we can arrange), the average molecular kinetic energy remains the same.

Thus $\Delta KE = 0$. Second, the gas is ideal, so there are no forces between molecules. There is therefore no potential energy involved, and $\Delta PE = 0$. (If the gas is real rather than ideal, there are weak attractive forces; expansion results in an increase in distance of separation of the molecules and a correponding increase in PE. For a real gas, ΔPE is small but > 0.) Thus the drive to minimum energy cannot be responsible for the tendency of a gas to expand. The only change on expansion is that each molecule has a larger volume available to it. The location of a molecule is less certain than it was before expansion. Expansion has the effect of increasing the disorder, or randomness, of the gas molecules.

What is the chance that the molecules in bulb B will spontaneously move through the stopcock into bulb A to give the state at the left of the figure? Such a process would lead to a decrease in the disorder of the system. But such a process has never been observed. Thus experience tells us that it is very improbable that the reverse process will occur. Expansion is spontaneous, meaning it occurs without outside assistance; compression is not spontaneous.

There is a natural direction for this process that leads to an increase in the randomness, or disorder, in the system. The amount of disorder in a system is measured by a quantity called entropy, symbolized S. The higher the disorder in a system, the higher the entropy. The change in entropy in a process is the difference between the entropies of the final and initial states, and is symbolized ΔS. It seems to be a law of nature that entropy increases ($\Delta S > 0$) in any spontaneous process.

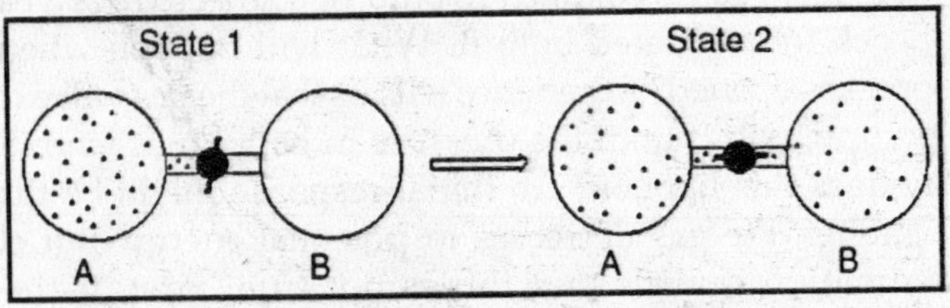

Fig. Ideal Gas is Confined in Bulb

A valid question at this point is this: "Why is there a tendency toward increased disorder?" A simple macroscopic example, based on the puzzle toy called Rubik's cube, will shed some light on this question. The toy consist of a cube with each face a different colour.

Each face can be independently rotated in either direction (clockwise or counterclockwise); rotation causes scrambling of the colours. We now consider what happens when we begin with the cube in the most ordered state — each face showing only one colour — and make a few random 90o face rotations. Begin with a 90° clockwise rotation of the top face of the cube. This produces a situation in which four of the faces show 2 colours.

If we agree that any configuration of the cube in which 2 or more faces show more than one colour is a disordered state, then our single rotation has produced a disordered state. Now give the cube to a blindfolded person. Inform the person that by making a single 90° rotation of one face, it is possible for him/her to return the cube to the ordered state.

Tell the person to make a single random move. What is the probability that this move will produce the ordered state? There are 12 possible 90o moves (6 faces, 2 directions per face). Of these, 1 will produce the ordered state, but 11 will produce other disordered states.

The probability that the person's move will produce the ordered state is 1 in 12. The probability that a random move will produce another disordered state is 11 times larger than the probability of producing the ordered state. The odds are very unfavorable, even when the cube is only one move away from the ordered state. If the cube were 6 moves away from the ordered state, the odds against restoring the ordered state by a sequence of 6 random moves would be 2,985,983 to 1. Disordered states are more probable than ordered ones.

Collections of molecules, like Rubik's cube, tend toward the most probable situation, because molecular motions are completely random. This requires only that some molecules in bulb A be moving toward the right.

On the other hand, once the gas has expanded, it can spontaneously contract only if, completely by chance, all of the 10^{23} molecules in bulb B happen to move toward the stopcock at the same time. The probability of this happening is essentially zero. The drive to maximum entropy is a consequence of the fact that a collection of particles, each moving randomly, is governed by the laws of statistics and probability.

We have indicated that the process is accompanied by increases in enthalpy and disorder. It is common knowledge that for any pure substance, the gas phase is favored at relatively high temperature and the solid phase at relatively low temperature. This simple, familiar fact allows us to conclude that law 2 is more important than law 1 at high T; and 1 is more important than 2 at low T. Although this statement has been based on a single simple process, it is generally true, for all systems. The question posed at the beginning of the section can now be addressed. Although the solid is the phase of minimum energy, the gas is the phase of maximum disorder. At low T, the drive to minimum energy predominates, and the solid is favored. At high T, the drive to maximum S predominates, and the gas is favored.

LIQUID-VAPOUR EQUILIBRIUM

Our observation following injection of water into the box was that the pressure rose gradually from zero to 23.8 torr at a temperature of 25°C, where it levelled off and remained constant indefinitely. The pressure resulted from formation of water vapour by evaporation of some of the liquid water.

We address two interesting questions:

1) Why does evaporation occur?
2) Why does the pressure rise to 23.8 torr, then stop changing?

The drive to minimum energy favors the liquid phase. However, the drive to increased disorder favors the gas phase. Two natural tendencies try to drive the process in equation in opposite directions. The result is a compromise in which some water is present as liquid and some as vapour. This is our macroscopic interpretation of the vapour pressure phenomenon. We now seek an understanding at the molecular level.

An important result from our models of the gas, liquid, and solid phases can be stated succinctly as follows: at a given temperature, all molecules have the same average kinetic energy, whether they are present in the solid, liquid, or gas phase. Further, the distribution of kinetic energies follows the Maxwell-Boltzmann distribution, regardless of phase. The general form of the Maxwell-Boltzmann distribution. In the ensuing discussion, we will apply this plot to the liquid phase of a pure substance.

As the curve indicates, at any moment there are some molecules moving very slowly, a large number moving with intermediate (near-average) kinetic energies, and a few molecules with very high kinetic energy. We focus now on the molecules near the surface of the liquid, because it is these that have a chance to escape from the liquid into the space above. Molecules near the surface possess a range of kinetic energies, like those in the bulk. A molecule near the surface and moving toward the surface will escape the liquid if its kinetic energy is sufficient to overcome the attractive forces of nearby molecules in the liquid. In other words, the kinetic energy must be at least equal in magnitude to the depth of the liquid potential well. We will call this minimum required kinetic energy the escape kinetic energy. The number of liquid molecules having at least this kinetic energy is proportional to the area under the curve to the right of a vertical line passing through the escape KE.

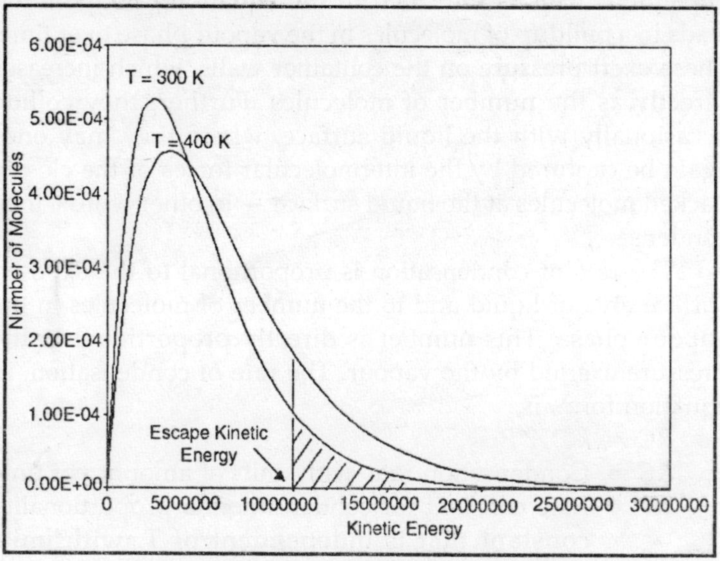

Fig. Maxwell-Boltzmann Distribution

When liquid water is injected into an evacuated box, it spreads out on the bottom and molecules begin to evaporate at a rate depending on two quantities. First, the number of molecules that evaporate per unit time is proportional to the number having at least the escape kinetic energy. Second, the evaporation rate is proportional to the number of molecules at the surface — i.e., to the exposed liquid surface area.

In equation form,

- RE = kE(T)*A
- RE = evaporation rate, with units of amount per time
- A = liquid surface area
- kE(T) = rate constant for evaporation, with units of amount per unit area per unit time.

(The constant, kE, is called a rate constant. It is a proportionality constant relating the evaporation rate to macroscopic variables on which rate depends. In this case, surface area is the only variable of importance, as long as temperature is constant. The size of the rate constant depends upon the fraction of molecules with KE greater than the escape KE. Because this depends on temperature, kE is temperature-dependent. This is indicated in the equation.) Evaporation leads to a buildup of molecules in the vapour phase over time. These exert pressure on the container walls, which increases directly as the number of molecules. Further, they collide occasionally with the liquid surface, where they may once again be captured by the intermolecular forces of the closely packed molecules at the liquid surface — in other words, they condense.

The rate of condensation is proportional to the exposed surface area of liquid and to the number of molecules in the vapour phase. This number is directly proportional to the pressure exerted by the vapour. The rate of condensation, in equation form, is,

$RC = kC*A*P$

RC = Condensation rate, with units of amount per time

kC = Rate constant for condensation, a proportionality constant that is independent of T, with units amount per time per area per unit pressure.

As the number of molecules in the vapour phase continues to increase as a result of a steady rate of evaporation, a point is eventually reached at which

$$RE = RC$$

Although both evaporation and condensation still occur, they occur at equal rates. There will be no further change in the number of molecules in the vapour phase, hence no further change in pressure. We describe this situation as a state of phase equilibrium.

$$l \Leftrightarrow v$$

The double arrow indicates that the two processes, evaporation and condensation, occur simultaneously and at equal rates. Phase equilibrium is dynamic, rather than static, because although it appears at the macroscopic level that nothing is happening (vapour pressure is constant), there is a lot of action at the molecular level.

This can be easily substantiated by starting with a system consisting of H_2O in the liquid phase and D_2O (deuterium oxide) in the vapour phase. After some time has elapsed, both forms of water will be found (by, for example, mass spectrometry) in both phases, indicating that evaporation and condensation are continuously occurring.

The pressure exerted by the vapour at equilibrium is called the equilibrium vapour pressure of the liquid, symbolized Pvap. Equations can be used to obtain an expression for Pvap in terms of the constants, kE and kC. Since at equilibrium, RE and RC are equal, we write

$$kE(T)A = kCAPvap$$

Solving for Pvap and cancelling the surface area terms in numerator and denominator gives equation

$$Pvap = kE(T)/kC$$

This equation shows that for a particular liquid at a fixed temperature, Pvap is constant, in agreement with experiment. The vapour pressure of a pure liquid provides the first example of an equilibrium constant, denoted Keq. For a liquid in equilibrium with vapour, Pvap = Keq. Keq is a ratio of rate constants.

Each liquid has a characteristic vapour pressure at a given T whose value depends primarily on kE. The greater the intermolecular forces, the greater the escape KE, and the smaller the value of kE. Small kE means smaller Pvap. Since the constant kC does not depend on the potential well depth, it is almost the same for all liquids. Equilibrium vapour pressures for several common laboratory liquids at 25oC are given in Table.

Table. Equilibrium Pvap at 25oC

Substance	Pvap (torr)
Water (H2O)	24
Ethanol (H2O)	65
Chloroform (H2O)	215
Diethylether (H2O)	545

Temperature Dependence of Vapour Pressure

When the temperature of a liquid-vapour equilibrium system is increased, the equilibrium vapour pressure increases. There are two reasons for this.The pressure due to the molecules already in the gas phase increases by the ideal gas law. This makes a small contribution to the vapour pressure increase. More molecules enter the gas phase. This is by far the larger contribution.

We now attempt to justify these statements at the molecular level. An increase in temperature means that the kinetic energy of the molecules increases in both phases. Molecules at the liquid surface jiggle more vigorously, and a greater number of them have sufficient KE to escape the potential well of the liquid and enter the vapour phase. The increase in the number of molecules in the vapour phase results in an increased vapour pressure.

As the curve shifts right with increasing T, the area under the curve to the right of the escape KE increases exponentially. Pvap therefore increases exponentially with T. Mathematically, the T-dependence of Pvap can be understood in terms of equation.

The minimum KE requirement for escape from the liquid is imbedded in the constant kE, which depends markedly on T. kC has almost no temperature dependence because no minimum KE is required for a molecule to fall into a potential well. The temperature dependence of Pvap therefore parallels that of kE , which directly reflects the shift in the Maxwell-Boltzmann distribution.

It can be shown, both experimentally and theoretically, that vapour pressure varies with temperature according to equation:

$$P_{vap} = e^{-\Delta Ho/RTe\Delta So/R}$$

This is usually written in logarithmic form as the Clausius-Clapeyron equation:

$$\ln P_{vap} = -\Delta H^o_{vap}/RT + \Delta S^o_{vap}/R$$

ΔH^o_{vap} and ΔS^o_{vap} are the standard molar enthalpy and entropy of vaporization, respectively. A plot of $\ln P_{vap}$ vs. 1/T should yield the enthalpy of vaporization, ΔH^o_{vap}, from the slope, and the entropy of vaporization, ΔS^o_{vap}, from the intercept.

This is a convenient experimental approach to the measurement of these quantities. The form of the Clausius-

Clapeyron equation is general in physical science; all molecular processes exhibit temperature dependence of this form.

The vapour pressure is plotted against temperature for several common laboratory liquids. The magnitude of P_{vap} for a liquid at a particular temperature and the rate of change of P_{vap} with temperature depend on the magnitude of the intermolecular forces in the liquid phase.

The stronger the forces, the lower P_{vap} at a given T and the steeper the rise in the curve. Of the liquids plotted water clearly has the largest intermolecular forces, due to strong hydrogen bonding. Having seen how P_{vap} depends on T, we consider a problem involving conceptual aspects of l-v phase equilibrium.

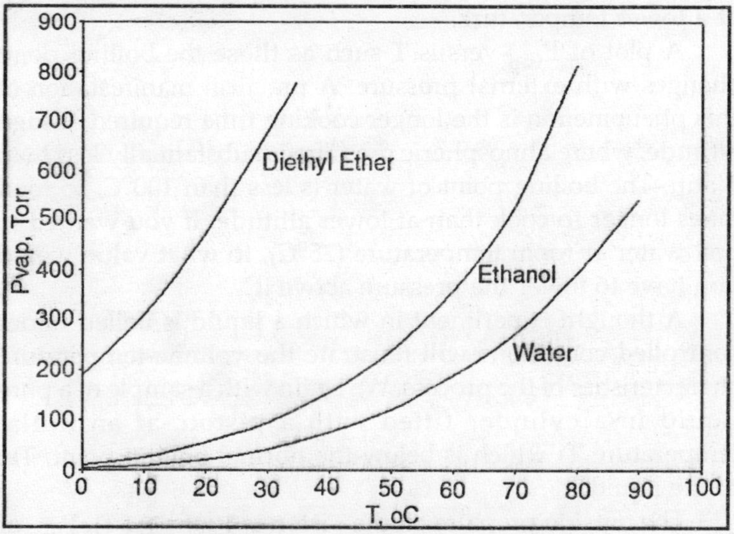

Fig. Vapour Pressure Versus Temperature

BOILING POINT

The boiling point of a liquid is the temperature at which the vapour pressure of the liquid is equal to the pressure above the liquid. The normal boiling point is the temperature at which the vapour pressure is 1 atm, the "normal" value of pressure above a liquid when it stands open to the atmosphere.

When a liquid boils, pockets of vapour form in the liquid and rise to its surface, where the vapour escapes to the atmosphere. The bubbles are not bubbles of air — they are pockets of the gaseous form of the substance. At temperatures where the vapour pressure of liquid is less than the pressure above the liquid, pockets cannot form because they are immediately collapsed by the greater outside pressure.

Only when the vapour pressure becomes equal to the outside pressure do the pockets become self-supporting. Under these conditions, boiling can occur. The relationship of the pressure inside a vapour pocket to the outside pressure.

The temperature at which a liquid boils can be lowered by reducing the pressure above it. Under these conditions the pockets become self-supporting at a lower value of P_{vap}, hence at a lower temperature.

A plot of P_{vap} versus T such as those the boiling point changes with external pressure. A practical manifestation of this phenomenon is the longer cooking time required at high altitude, where atmospheric pressure is substantially less than 1 atm. The boiling point of water is less than 100°C, so food takes longer to cook than at lower altitude. If you wanted to boil water at room temperature (25°C), to what value would you have to lower the pressure above it?

A thought experiment in which a liquid is boiled under controlled conditions will illustrate the volume-temperature characteristics of the process. We begin with a sample of a pure liquid in a cylinder fitted with a piston, at an initial temperature Ti which is below the normal boiling point, Tb, of the liquid.

The outside pressure on the piston is 1 atm. At Ti, P_{vap} of the liquid is less than 1 atm so vapour cannot exist in the cylinder. We justify this statement as follows. Suppose there were some vapour above the liquid in the cylinder. It would exert a pressure P_{vap} on the piston.

Because $P_{vap} < 1$ atm, the piston would begin to move in. In response, vapour would condense to maintain the equilibrium value of P_{vap}. This would cause the piston to move further in.

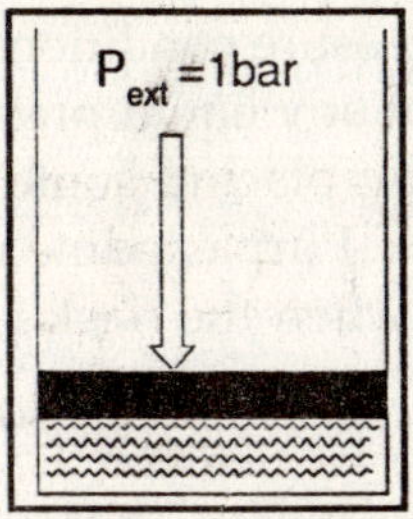

Fig. Liquid in Crystal/Piston

This process of vapour condensation and inward movement of the piston would continue until there was no vapour left. Thus in the beginning, there can be no vapour present. We now slowly heat the liquid to its normal boiling point, and boil it. As long as T < Tb, P_{vap} < 1 atm and the piston will not move. But at the T where P_{vap} becomes equal to 1 atm, vapour begins to form in the cylinder, and the piston begins to recede. More and more liquid evaporates in order to maintain P_{vap} at 1 atm until all of the liquid disappears. At this point only vapour will be present in the cylinder. Finally, we continue heating the vapour. The volume of vapour increases with T according to the ideal gas law. A plot of the volume of the system as a function of T for the boiling process. Several features of the plot are important:

At T > Ti but < Tb, the system volume remains essentially constant and small, because in this T range no vapour exists in the system, and the volume of the liquid does not vary much with T. Once T reaches Tb, the liquid begins to boil. Throughout the boiling process, T remains constant because all added heat is used to overcome the intermolecular forces in the liquid, increasing the PE of the molecules, and to push back the atmosphere. The KE of the molecules does not change.

When all liquid has evaporated, added heat is used to increase the KE of the vapour molecules. T and V increase in a manner consistent with the ideal gas law.

SOLID-VAPOUR EQUILIBRIUM

Just as liquid and vapour can coexist in equilibrium under

certain temperature-pressure conditions, so too can solid and vapour coexist, without any liquid present. This situation can be directly observed by placing some solid iodine (I2) in a closed Erlenmeyer flask. Purple iodine vapour is visible in the space above the solid when the flask is held against a white background.

$$s \Leftrightarrow v$$

The double arrow means that conversion of solid to vapour (sublimation) and vapour to solid (deposition) occur at equal rates, so that there is no macroscopic change in the system.

The temperature dependence of the pressure of vapour in equilibrium with solid is similar to that of the liquid vapour pressure. It is governed by a form of the Clausius-Clapeyron equation in which the enthalpy and entropy of vaporization are replaced by the corresponding quantities for the sublimation process, $\Delta H^o{}_{sub}$ and $\Delta S^o{}_{sub}$:

$$\ln P_{vap} = -\Delta H^o{}_{sub}/RT + \Delta S^o{}_{sub}/R$$

As shown in the figure, the straight line for the solid has a more negative slope and a more positive intercept than the liquid line. This is true because

$$\Delta H^o{}_{sub} > \Delta H^o{}_{vap}$$
$$\Delta S^o{}_{sub} > \Delta S^o{}_{vap}$$

The increase in potential energy for conversion of solid to vapour must exceed that for conversion of liquid to vapour, and similarly for the increase in disorder, because the solid is the phase of lowest PE and disorder. The inequalities that the two lines cross at some temperature, Tt. At T > Tt, the vapour pressure of the solid exceeds that of the liquid; at T < Tt, the vapour pressure of the solid is less than that of the liquid.

Ultimately we would like to develop a graphical presentation of the system that shows directly the vapour pressure of the system as a function of temperature. To do this, we translate the graph of ln P_{vap} versus 1/T to one in which P_{vap} is plotted directly against temperature. Each possible state of the system is represented by a (P,T) point on this plot. To perform the translation, we make use of 3 facts:

ln P_{vap} for the solid is less than ln Pvap for the liquid at high 1/T (low T). Therefore P_{vap} for the solid is less than P_{vap} for the liquid at low T.

The Clausius-Clapeyron equation for the solid has a steeper slope than that for the liquid. Therefore the vapour pressure of the solid rises more rapidly with T than does that for the liquid.

As we have already seen, plots of P_{vap} versus T curve exponentially upward.

The curve labelled "l-v" represents all P-T combinations where liquid and vapour can coexist in equilibrium. The curve labelled "s-v" represents all P-T combinations at which solid and vapour can coexist in equilibrium. Because the point at which these two curves intersect lies on both curves simultaneously, all three phases must coexist together here. This is called the triple point, with temperature Tt.

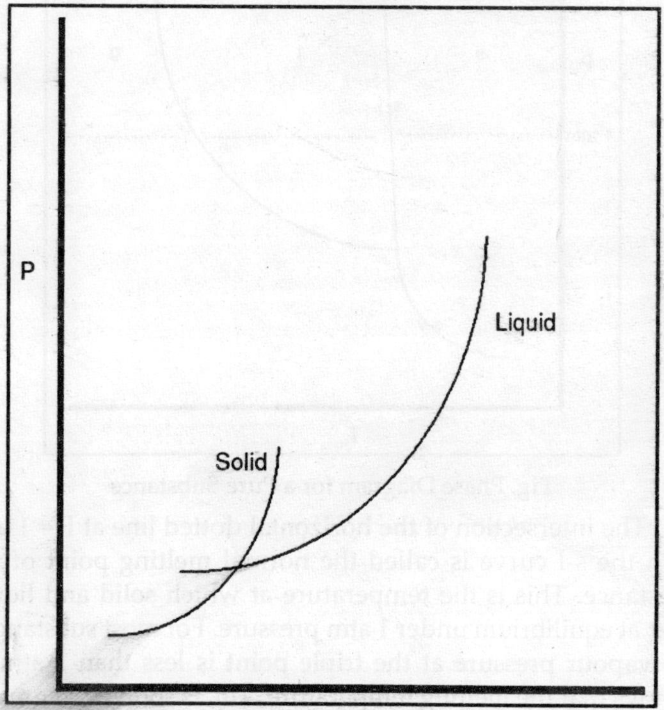

Fig. Vapour Pressure Temperature Curves for Solid and Liquid

SOLID-LIQUID EQUILIBRIUM

Completion of our graphical presentation of the states of the system requires only the addition of a curve of all P,T combinations at which solid and liquid coexist in equilibrium. One point of this curve — the triple point — is already plotted. For most substances, the "s-l" line slopes sharply up from the triple point, with a positive slope. The steep slope of this line reflects the fact that solid and liquid are both condensed phases, with properties that depend very little on pressure. A big change in pressure results in only a small change in the temperature of the equilibrium system.

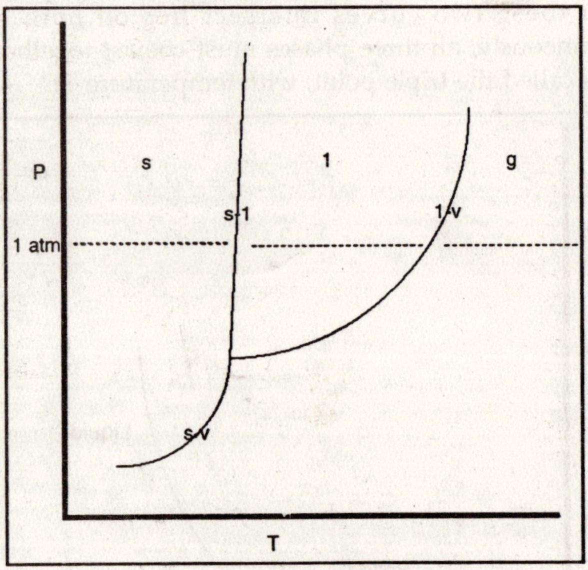

Fig. Phase Diagram for a Pure Substance

The intersection of the horizontal dotted line at P = 1 atm with the s-l curve is called the normal melting point of the substance. This is the temperature at which solid and liquid exist at equilibrium under 1 atm pressure. For most substances, the vapour pressure at the triple point is less than 1 atm. It follows that the melting temperature, Tm, is somewhat greater than the triple point temperature, Tt (although the difference is usually slight). The intersection with the l-v curve is the normal boiling point of the substance. The temperature at which the vapour pressure of the liquid reaches 1 atm.

PHASE DIAGRAM FOR A PURE SUBSTANCE

Although details of temperature and pressure vary from one substance to another, all pure substances have phase diagrams that resemble the generalized diagram in the figure. A phase diagram is a superposition of l-v, s-v, and l-s curves for a substance, and shows the phase behaviour of the substance as a function of temperature and the pressure to which the substance is subjected. It is important to realise that the diagram is strictly applicable only to a sample of substance contained in a closed system in which no other substance is present. A convenient apparatus in which to study the phase behaviour of a substance under these restrictive conditions is a cylinder fitted with a piston (e.g., a syringe). Such an apparatus allows the system to be maintained at any desired temperature, and the desired pressure may be exerted on the substance via the piston. In discussing phase diagrams, we assume that the pure substance being studied is enclosed in such an apparatus.

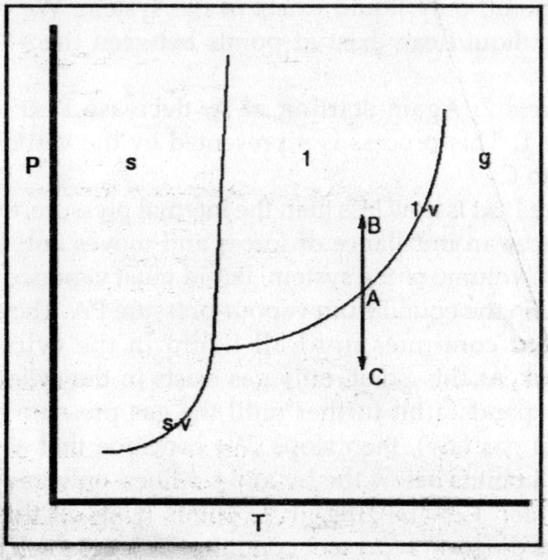

Fig. Effect of pressure changes on phases of a pure substance

Points that lie on the lines in a phase diagram represent those P-T conditions at which two phases coexist in equilibrium. What is the significance of (P,T) points that lie off the lines? As agreed above, the substance is confined in a cylinder/piston apparatus, shown at top. The phase diagram for the substance is shown below. Initially the system is at point A in the diagram. Consistent with this, the cylinder contains the liquid and vapour phases in equilibrium at TA and vapour pressure PA. The external pressure on the piston is also kept at PA, so the piston is stationary. Starting at A, we carry out two processes.

Process 1: Increase the external pressure to PB, while keeping temperature constant at TA. This process is represented by the vertical arrow from A to B in the figure.

Since Pext (equal to PB) now exceeds the internal pressure, PA, the piston moves in. Vapour in the system condenses to maintain the equilibrium vapour pressure, PA. The piston continues to move in until all vapour condenses and the piston rests directly on top of the liquid. When the vapour is completely condensed, the internal pressure rises to PB, and the piston stops. The system is now at point B on the diagram. At this point, only liquid exists in the system. We conclude that only liquid can exist at points between the s-l and l-v curves.

Process 2: Again starting at A, decrease Pext to PC at constant T. This process is represented by the vertical arrow from A to C.

Since Pext is now less than the internal pressure, the piston experiences an imbalance of forces and moves out. With the increased volume of the system, liquid must vaporize in order to maintain the equilibrium vapour pressure PA. The outward movement continues until all liquid in the cylinder has vaporized. At this point, only gas exists in the cylinder. The piston expands a bit further until the gas pressure drops to PC (ideal gas law), then stops. We conclude that at point C and at all points below the l-v and s-v lines, only gas exists in the cylinder. Generalizing, (P,T) points lying off the lines in the phase diagram represent conditions under which only one phase may be present at equilibrium. Regions in which only one phase can exist are indicated by the appropriate symbols in the diagram. Notice that the solid phase is stable at relatively low temperatures and high pressures; the gas phase is stable at relatively high temperatures and low pressures; and the liquid is stable at intermediate T and P.

LE CHATELIER'S PRINCIPLE

Le Chatelier's principle is an important tool for predicting the effects of pressure and temperature changes on phase equilibria: A system in equilibrium subjected to an external stress responds to reduce the effect of the stress. There are two stresses that are important for phase equilibria. We examine them in terms of the stress principle. Change in pressure. Suppose we increase the pressure on a system in phase equilibrium, say a liquid-vapour equilibrium.

$$l \Leftrightarrow v$$

Le Chatelier's Principle predicts that the equllibrium will shift to offset the effect of the stress. Intuitively, we expect the system to decrease its volume, to retreat from the increased pressure. This will happen in a liquid-vapour equilibrium system if some vapour converts to liquid. The system shifts toward the denser phase.

A liquid is in equilibrium with vapour at left. The piston is at position A, and the external pressure is equal to the equilibrium vapour pressure of the system. The piston is then suddenly moved down to a new position, B. The instantaneous effect of this stress is to increase the pressure exerted by the gas inside the cylinder. The increase in pressure causes the rate of condensation, RC, to become greater than the rate of evaporation, RE, which depends only on temperature. A net flow of molecules from vapour to liquid results — i.e., the system shifts toward the denser liquid phase — to reduce the internal pressure to the equilibrium value. Once this value is reached, RE and RC will again be equal, and the system will be in equilibrium again. The difference between the final and initial equilibrium states is that there is less vapour and more liquid in the final state.

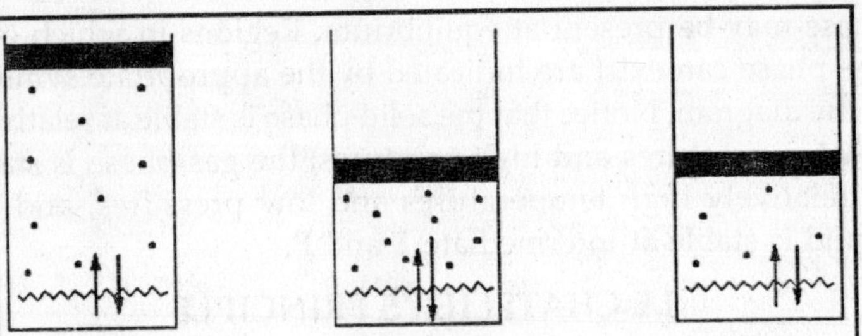

Fig. Effect of Volume Decrease on Liquid Vapour Phase Equilibrium

Change in temperature. Le Chatelier's Principle predicts that increasing the temperature of a system in phase equilibrium will cause it to shift to use heat. Conversely, a temperature decrease will result in a shift that produces heat. An increase in T will cause a shift toward the vapour phase (to the right), since conversion of liquid to vapour is endothermic and consumes heat. This will increase the vapour pressure of the liquid, as we already know.

$$l \Leftrightarrow s$$

For the process as written, H < 0. An increase in the temperature of the system causes a shift in the endothermic direction; i.e., to the left.

PHASE DIAGRAM FOR WATER

Water, a common substance upon which we depend for life, has unusual phase behaviour. If water initially at 25oC is cooled to its freezing point of 0°C, it becomes gradually denser until a temperature of 4°C is reached. At this temperature, water has its maximum density. With further cooling, the density of the liquid decreases again until solid (ice) begins to form. Surprisingly, liquid water near 0°C is denser than solid water. Thus water is unusual, since for most pure substances, the solid phase is denser than the liquid. The low density of ice is a consequence of a very open lattice structure that results from extensive hydrogen bonding between neighboring molecules. Each water molecule in the lattice is engaged in 4 hydrogen bonds, 2 using the O atom, and 1 for each of the H atoms. Breakdown of this hydrogen bonded network upon melting results in a collapse of the open lattice, with an increase in density.

What will happen if pressure is applied to a block of ice in equilibrium with liquid at 0°C? According to Le Chatelier, the solid-liquid equilibrium will shift toward the denser phase, in this case the liquid. Application of pressure to ice melts it. In terms of the phase diagram, this can happen only if the l-s equilibrium line has a negative slope, rather than the positive slope of most substances. The slope of the s-l line therefore gives information about the relative densities of solid and liquid phases.

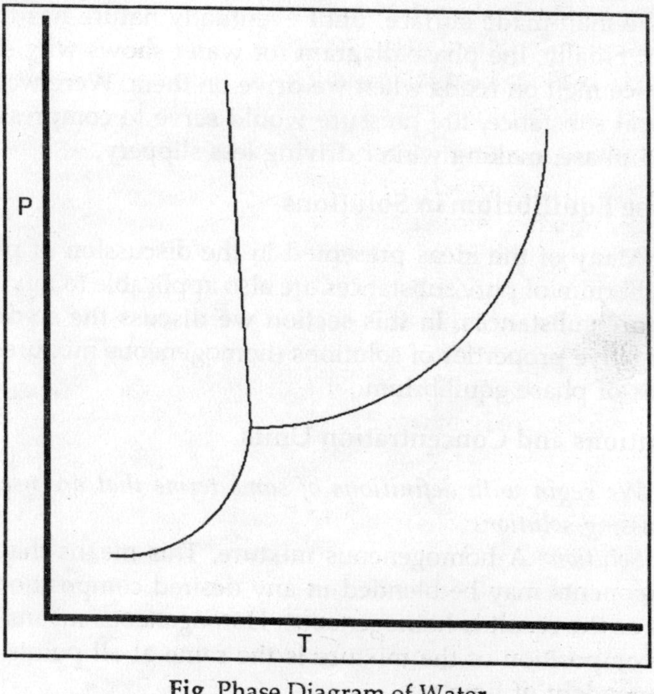

Fig. Phase Diagram of Water

There are a number of practical manifestations of the unusual phase diagram for water. Perhaps the most important is that lakes and other natural bodies of water freeze from the top down, because as water cools from 4 to 0°C , it becomes less dense and rises to the surface. The layer of ice that forms on the surface then insulates the water below and prevents the entire lake from freezing. Aquatic life is thus enabled to survive the cold of winter. A second, and costly, consequence of the reversed s-l line is the damage done to roads by the freezing of water in surface cracks.

The water freezes, expands, and pushes the pavement out of the way. The result is frost heaves and potholes. The same process is responsible for the erosion of natural rock formations and for the gradual recovery by nature of man-made structures. In the latter case, water freezes in cracks and fissures, opening them up to the growth of vegetation.

The root systems of the vegetation cause further breakup of the man-made surface, until eventually nature has taken over. Finally, the phase diagram for water shows why snow and ice melt on roads when we drive on them. Were water a typical substance, tire pressure would serve to compress the solid phase, making winter driving less slippery.

Phase Equilibrium in Solutions

Many of the ideas presented in the discussion of phase equilibrium of pure substances are also applicable to mixtures of pure substances. In this section we discuss the so-called colligative properties of solutions (homogeneous mixtures) in terms of phase equilibrium.

Solutions and Concentration Units

We begin with definitions of some terms that are used in discussing solutions:

Solution: A homogeneous mixture. This means that the components may be blended in any desired composition, as long as the result is homogeneous. Homogeneous means that the composition of the mixture is the same at all points and independent of time.

Solvent: The component present in major amount. In most cases, the solvent is a liquid.

Solute: A component present in minor amount. In general a solution may contain more than one solute.

If we add 1.0 g of sugar to 10 mL of water in a beaker, the sugar falls to the bottom of the beaker where it forms a pile. Over the period of a few minutes, the pile becomes smaller until finally no solid sugar remains visible. The sugar has dissolved in the water. The result is a solution of sugar in water. Because water is present in major amount, it is the solvent, and sugar is the solute. The solution is clear and displays all of the usual properties of a liquid. There is no cloudiness (opacity) to indicate the presence of the sugar.

In preparing a solution, it is usually important to specify, as precisely as possible, the amount of solute present per unit amount of solvent or per unit amount of solution. This is called the concentration of the solution. The concentration is an intensive property of the solution, independent of the amount of solution present. In this respect, concentration is similar to density. There are several ways to express concentration, discussed below.

Molarity, M. If A represents the solute, the molarity of the solution is the number of moles of A per liter of solution. Symbolically,

MA = [A] = moles A/1 L solution

Molality, m. The molality of a solution of A in some solvent is the number of moles of A per kilogram of solvent. Symbolically,

MA = moles A/1 kg solvent

Mole fraction, X. The mole fraction of A in a solution containing A and at least one other substance is the ratio of the number of moles of A to the total number of moles of all substances. Symbolically,

XA = moles A/total moles

If A is the only solute, then XA = moles A/(moles A + moles solvent)

Suppose that we require 100 mL of a 0.10 M sulfuric acid solution, but all we have available is commercial concentrated sulfuric acid, which is 18 M in H_2SO_4. We can prepare the required solution by dilution. Dilution is the process of adding a known volume of a concentrated solution of a reagent to water (or another solvent) to give a less concentrated solution.

It is a very frequently-used laboratory technique. The key to dilution is that the number of moles of H_2SO_4 in the required volume of 18 M H_2SO_4 is the same as the number of moles in 100 mL of 0.1 M H_2SO_4. The number of moles is the product of solution volume and molarity both before and after dilution. Using subscript "i" for initial quantities and "f" for final quantities,

$$V_iM_i = V_fM_f$$

Here V_i is the volume of concentrated H_2SO_4, M_i is its molarity, V_f is the volume of dilute H_2SO_4, and M_f is its molarity. V_i can easily be calculated from the known values of the other 3 quantities. The result is 0.556 mL.

Solutions have some interesting properties called colligative properties. They depend on the nature of the solvent and the amount (but not the identity) of the solute. These properties may be understood in terms of phase equilibrium. We begin with the effect of a solute on the vapour pressure of a liquid solvent.

RAOULT'S LAW

Consider a solution of a non-volatile solute, B, in a volatile solvent, A. A volatile substance has an appreciable vapour pressure at room temperature; a non-volatile substance has essentially zero vapour pressure. How is Pvap of the solvent in the solution related to Pvap of the pure solvent?

$$P_A = X_AP_A^o$$

P_A = the vapour pressure above the solution due to solvent, X_A = the mole fraction of solvent, and P_A^o = the vapour pressure of pure solvent. This equation is called Raoult's Law. We can make several statements about this law.

- $X_A < 1$ so the vapour pressure of the solution is less than that of pure solvent.
- No term in the equation involves any chemical or physical property of the solute. It doesn't matter what the solute is, only that it is non-volatile.
- The equation is ideal because it holds only for ideal solutions. An ideal solution is one in which intermolecular forces between A molecules, between

B molecules, and between A and B molecules are the same. If these forces are the same, there is no enthalpy change when solute dissolves in solvent. It is important to realise that Raoult's Law has very limited applicability. Very few real solutions even approach the ideal, and most are definitely NOT ideal. These do not obey Raoult's Law. Some solutions are nearly ideal when very dilute; thus Raoult's law approximately describes the vapour pressures above most dilute solutions.

- The law is purely statistical. It says that the tendency for solvent molecules to escape the liquid is in proportion to the relative abundance (mole fraction) of solvent molecules at the liquid surface.

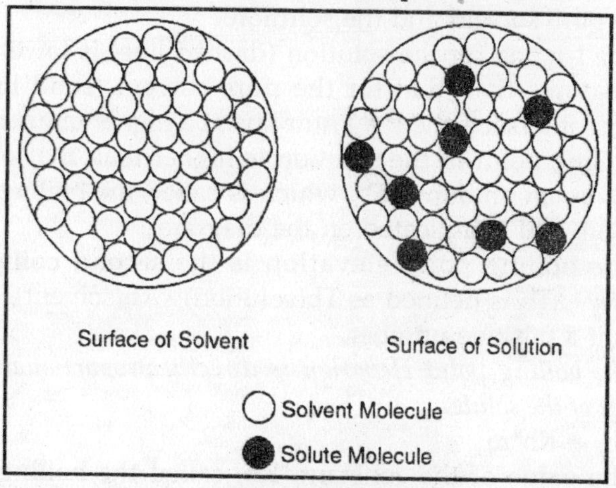

Fig. Raoult's Law

Raoult's Law can be understood physically by considering the liquid surfaces of pure solvent and solution. In pure solvent, all surface molecules are solvent. However, in the solution, a fraction of molecules at the surface are solute molecules, which are non-volatile (cannot escape to the vapour phase). These reduce the surface area available for escape of solvent molecules, so that the evaporation rate of solvent is reduced from that in pure solvent. The fraction of surface area occupied by solvent molecules is directly related to the mole fraction of solvent. The evaporation rate, and consequently the vapour pressure, is reduced by a factor equal to the ratio of surface areas, which is the mole fraction of solvent.

COLLIGATIVE PROPERTIES

The extent to which the vapour pressure is lowered depends quantitatively on the amount of solute present. Defining $\Delta P = P_A{}^o - P_A$, and substituting for P_A using Raoult's Law, gives,

$\Delta P = P_A{}^o - X_A P_A{}^o = P_A{}^o(1 - X_A) = X_B P_A{}^o$ where XB is the mole fraction of solute.

The vapour pressure lowering, ΔP, is the first of four colligative properties. The remaining three are consequences of ΔP, and may be examined in terms of the phase diagrams for the pure solvent and the solution.

The l-v line for the solution (dashed line) is lower at all temperatures than that for the pure solvent (solid line). It therefore intersects the P = 1 atm line at a higher temperature. The boiling point of the solution is higher than that of pure solvent by an amount ΔTb, which is called the boiling point elevation, and is indicated on the diagram.

The boiling point elevation is the second colligative property. ΔTb is defined as Tb(solution) - Tb(solvent), which makes it a positive number.

The boiling point elevation is directly proportional to the molality of the solute:

Tb = Kb*m

The value of the constant, Kb, called the boiling point elevation constant for the solvent, depends only on properties of the solvent.

Because the l-v line for the solution is lower, it intersects the s-v line at lower T than does the l-v curve for the pure solvent. The s-l line for the solution therefore lies to the left of that for the pure solvent.

Consequently the triple point for the solution is lower than that for the solvent by an amount ΔTf, indicated on the diagram. Over small ranges of T and P, the s-l lines are linear and roughly parallel.

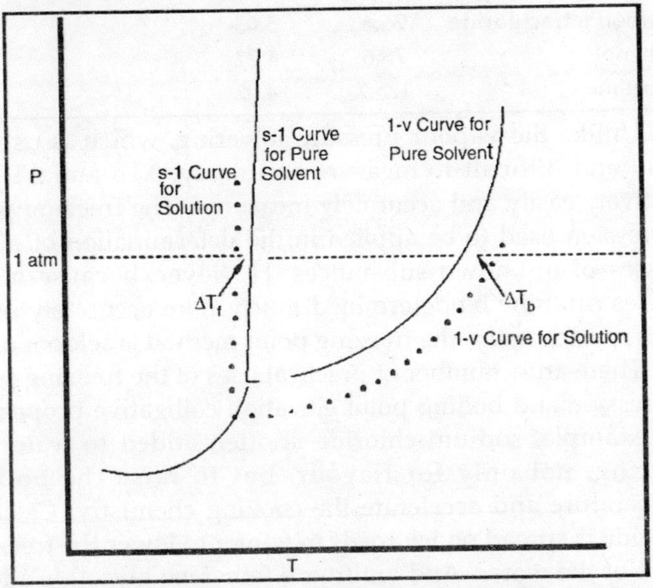

Fig. Phase Diagram of Pure Substance and a Solution for which the Substance is the Solvent

It follows that the freezing point for the solution is lower than that for the solvent by about the same amount, ΔTf. ΔTf is called the freezing point depression, and is defined to be a positive number: ΔTf = Tf - Tf(solution). It is the third colligative property. The freezing point depression is similar to the boiling point elevation in being directly proportional to the solute molality:

ΔTf = Kf*m

Kf is the freezing point depression constant and depends only on properties of the solvent. Table gives values of Kb and Kf for several common solvents.

Table. Molal Boiling Point and Freezing Point Constants

Solvent	Tb(oC)	Kb	Tf(oC)	Kf
Acetic acid	118.2	2.93	17	3.90
Chloroform	61.2	3.63		
Naphthalene			80	6.8
Water	100	0.52	0	1.86
Camphor			179	40

Carbon tetrachloride	76.6	5.03
Ethanol	78.6	1.22
n-octane	125.7	4.02

Unlike the vapour pressure lowering, which is usually small and difficult to measure accurately, ΔTb and ΔTf are relatively easily and accurately measured. The freezing point depression used to be applied in the determination of molar masses of unknown substances. However, because molar masses can now be determined much more accurately using mass spectrometry, the freezing point method is seldom used.

There are a number of practical uses of the freezing point depression and boiling point elevation colligative properties. For example, sodium chloride is often added to water for cooking, not only for flavour, but to raise the boiling temperature and accelerate the cooking chemistry. Calcium chloride is spread on icy roads in winter to lower the freezing point of the water. And antifreeze (ethylene glycol) is added to the cooling systems of automobiles to protect against boilover in the summer and coolant freeze in winter.

The fourth colligative property is called osmotic pressure. It can be understood in terms of the apparatus which consists of two arms, separated by a semi-permeable membrane. This is a membrane with pores that allow passage of small molecules (solvent) but block larger molecules (solute). In the left arm of the apparatus is placed a pure solvent, usually water. In the right arm is placed a solution of the same solvent. The dots in the figure represent molecules of solute present in the solution. Solvent molecules can pass through the membrane in either direction. However, their passage from right to left (from solution to pure solvent) is impeded somewhat because solute molecules block their access to some of the pores. More solvent molecules per unit time pass from solvent to solution than in the reverse direction. The result is that the amount of solvent in the right arm increases with time and the solution becomes more dilute. It is possible to prevent the net flow of solvent from left to right by applying pressure to the top of the liquid column in the right arm of the apparatus. The applied pressure increases the rate of passage of solvent from solution to solvent until it is equal to the rate of passage in the other direction. Under these conditions, the system is in equilibrium and the concentration of the solution does not change with time. The pressure that must be applied to stop the net flow of solvent is called the osmotic pressure, symbolized. Its magnitude is directly proportional to the temperature and the molarity of the solution. The ideal gas constant, R, is the proportionality constant:

K= M*R*T

Realizing that molarity is moles of solute per unit volume of solution, expressed in liters to a form analogous to the ideal gas law:

PV= n*R*T

It may seem odd that the ideal gas constant appears in an equation that describes a property of a liquid solution. However, a non-volatile solute dissolved in a solvent behaves much like a gas, because the solute molecules are far apart and are free to roam over the entire solution volume. This similarity between a gas and a dissolved solute causes osmotic pressure to obey a gas-law type equation. The similarity in fact extends further. The same value as the pressure of a sample of gas occupying the same volume at the same temperature. We know that the gas will expand if the pressure exerted on it is less than the pressure that it exerts.

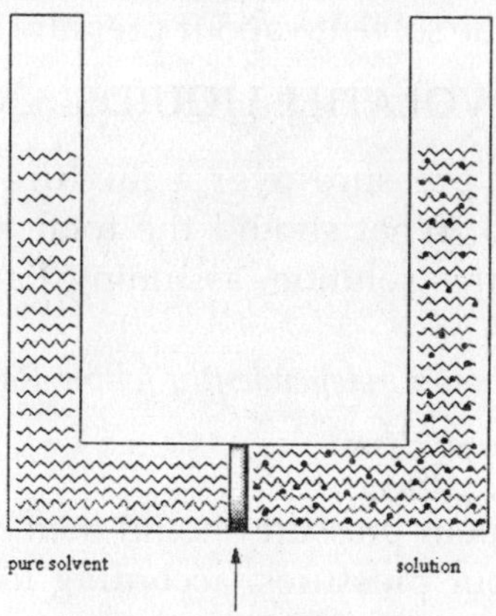

Fig. Osmotic Pressure

A simple calculation using equation that concentrations as low as 10^{-4} M give readily measureable osmotic pressure. The osmotic pressure is therefore much more sensitive to the presence of solute than the other colligative properties.

It can be used to determine the molar masses of huge molecules like proteins, which dissolve in water to give solutions that may be concentrated in terms of weight per cent, but are of very small molarity due to the large molar masses (on the order of 10000 m). It may not be clear that osmotic pressure, the tendency for solvent to flow spontaneously from solvent to solution across the barrier, is a consequence of the vapour pressure lowering of the solvent in the solution. Two beakers, one containing pure solvent, the other the same solution that is in the right arm of the apparatus.

The beakers are placed in a closed container. Solvent begins to evaporate from each beaker in an attempt to establish the equilibrium vapour pressure. However, this pressure is less for the solution than for the pure solvent.

As pure solvent evaporates from the left beaker in an attempt to establish the vapour pressure, Povap, it condenses in the right beaker as the solution tries to lower the vapour pressure to its equilibrium value, P_{vap}.

This process has the same result as the process in the solution gradually becomes more dilute due to spontaneous "flow" of solvent into it. The origin of the osmotic pressure phenomenon in the solvent vapour pressure lowering is clear.

MIXTURES OF VOLATILE LIQUIDS

The vapour pressure over a mixture of two volatile liquids, A and B. What should the total vapour pressure, PTvap, be over this solution, assuming Raoult's Law to be valid?

Each liquid should independently follow Raoult's Law:

$P_{vap}(A) = P^{o}_{vap}(A)X_A$

$P_{vap}(B) = P^{o}_{vap}(B)X_B$

The total vapour pressure should then be the sum of the two partial vapour pressures, according to Dalton's partial pressure law:

$$PT_{vap} = P_{vap}(A) + P_{vap}(B) = P^o_{vap}(A)X_A + P^o_{vap}(B)X_B$$

Plots of $P_{vap}(A)$, $P_{vap}(B)$, and PT_{vap} versus XA should be linear. If the intermolecular force between an A and a B molecule is identical to that between two A molecules or two B molecules, the solution is ideal and produces plots like those. A mixture of hexane (a chain of 6 carbon atoms with attached hydrogens) and heptane (a chain of 7 carbon atoms with attached hydrogens), which have very similar molecular structures and compatible intermolecular forces, is very nearly ideal. The same is true of a mixture of carbon tetrachloride and carbon tetrabromide.

For a pair of ideal liquids, there is essentially no change in the intermolecular forces on mixing. Thus there is no change in potential energy; the enthalpy change accompanying mixing, ΔH_{soln}, is zero. The volume change on mixing, ΔV_{soln}, is nearly zero as well, because intermolecular separation distances will not change unless there is an increase or decrease in forces of attraction.

Ideal liquid mixtures are rare. It is much more common for two liquids to form a non-ideal mixture, for which the vapour pressure plots are non-linear. Mixtures producing plots like this are said to exhibit positive deviations from Raoult's Law, because the vapour pressures are larger than expected for an ideal mixture.

Larger vapour pressures imply that the molecules more readily escape from the liquid phase than they would if the intermolecular forces were ideally compatible. Addition of heptane to acetone disrupts the relatively strong dipole-dipole forces between acetone molecules. The weaker dipole-induced dipole forces between acetone and heptane molecules are insufficient to make up for the disruption. Acetone molecules therefore escape the liquid more easily when surrounded by heptane molecules than when surrounded by other acetone molecules.

When intermolecular forces between unlike molecules are weaker than those between like molecules, positive deviations are observed. H_{soln} and ΔV_{soln} for such mixtures are both greater than zero.

In this case, the vapour pressures are smaller than the ideal values, implying that the molecules escape the liquid phase less readily in the mixtures than they would if the intermolecular forces were ideally compatible. Plots like those in exhibit negative deviations from Raoult's Law. Intermolecular forces between acetone and chloroform molecules are relatively strong, involving some hydrogen bonding in addition to the expected dipole-dipole contribution.

A molecule of chloroform therefore escapes the liquid less readily when surrounded by acetone molecules than when surrounded by other chloroform molecules. When intermolecular forces between unlike molecules are stronger than those between like molecules, negative deviations are observed. ΔHsoln and ΔVsoln for such mixtures are both less than zero.

To some extent it is possible to predict whether or not two liquids will show deviations from Raoult's Law based on the compatibility of their molecular structures. Two liquids with very similar, non-polar molecular structures tend to form nearly ideal solutions, with at most only small deviations from linearity.

The hexane/heptane and CCl_4/CBr_4 examples cited above illustrate this. Solutions of liquids with similar polar structures often exhibit negative deviations from Raoult's Law. In contrast, two liquids with quite different structures, one polar and one non-polar, are incompatible, and therefore show positive deviations.

When the molecular structures of the liquids are very different, as is the case for H_2O (polar) and CCl_4 (non-polar), the molecules are so eager to escape one another that they do so even in the liquid phase. Such liquids do not dissolve in one another, and are said to be immiscible (i.e., do not mix). Immiscible liquids form two layers when mixed, with the denser liquid on the bottom.

Additional Equilibria Involving Two Phases

In our discussion of the phases of a single pure substance, we found that equilibrium involving two phases is achieved when the rate of passage from phase 1 to phase 2 is balanced by an equal rate of passage from phase 2 to phase 1:

Rate1→2 = Rate2 → 1

The rate of passage from phase 1 to phase 2 is directly proportional to the concentration of substance in phase 1 and to the surface area, SA, of contact between the two phases. A similar statement holds for the rate of passage from phase 2 to 1:

Rate1 → 2 = k1 → 2*SA*[substance]1

Rate2 → 1 = k2 → 1*SA*[substance]2,

When these two rates become equal, equilibrium exists, and the following relationship characterizes it:

Rate1 → 2 = Rate2 → 1

k1 → 2*SA*[substance]1 = k2 → 1*SA*[substance]2

Keq = k1 → 2/k2 → 1 = [substance]2/[substance]1

Keq is an equilibrium constant, and is seen to be the ratio of rate constants for opposing processes. We now consider a number of other two-phase equilibrium situations for which the above considerations apply.

SOLUTE

When two immiscible liquids such as water and chloroform ($CHCl_3$) are placed in contact, they mutually exclude each other, forming two liquid layers. The less dense liquid floats on the denser one, and a boundary surface is clearly visible. It is interesting to think about what will happen if a third substance, solid or liquid, is added to the system consisting of two immiscible liquids, A and B. If the third substance, C, is soluble in one of the liquids and insoluble in the other (B), it will dissolve exclusively in the A layer, as in example. More interesting is the situation in which C is soluble in both liquids to at least some extent. In this case, we find that C partitions, or distributes, itself between the two liquid layers. We further find that the concentration of the third substance is higher in the liquid with which its intermolecular forces are most compatible. The ratio of the concentration of C in liquid B to that in liquid A is found to attain a constant value Kd:

Kd = [C]B/[C]A

This constant ratio is established no matter how the 3-component mixture is prepared. Three such ways follow:

A and B are first brought into contact. C is then added, and the container is shaken.

C is first dissolved in A. A (containing C) and B are then brought into contact, and the container is shaken.

C is first dissolved in B. A and B (containing C) are then brought into contact, and the container is shaken.

In addition, the ratio is reestablished if the mixture is perturbed in any of the following ways:

More A is added.

More B is added.

More C is added.

We can understand this behaviour in terms of the diagram, which shows layers of A and B in contact at a boundary, and molecules of C initially dissolved in liquid A. The boundary between the two liquids provides a "doorway" from one to the other through which C molecules, which have reasonable interactions with both A and B molecules, can pass. Thus molecules of C will tend to move across the boundary from liquid A, where they are initially dissolved, into liquid B, in which there are initially no molecules of C. In other words, they begin to distribute themselves between the two liquids. After some time, enough C molecules will have passed into liquid B that the rates of passage of C from A to B and back will be the same.

C dissolved in A ⇔ C dissolved in B

RateA → B = RateB → A

The rate of passage from A to B is directly proportional to the concentration of C in liquid A, and to the surface area over which A and B are in contact.

RateA → B = kA → B*SA*[C]A

Similarly,

RateB → A = kB → A*SA*[C]B

Equating the two rates at equilibrium gives

[C]B/[C]A = kA → B/kB → A = Kd

Kd is an equilibrium constant, the ratio of rate constants

for opposing processes. Once the equilibrium situation is attained, there will be no further change in the concentration of C in the two layers. In other words, the constant ratio in will be attained. From the magnitude of Kd, we may easily deduce which liquid C prefers. If Kd < 1, C prefers A; if > 1, C prefers B.

In summary, distribution of a substance between two liquid phases is an example of dynamic equilibrium, a situation in which two opposing processes (here, movement of C both ways across the boundary) occur at equal rates.

Solutions of Molecular Solids in Liquids

A molecular solid is one in which covalent molecules are arranged in a lattice held together by intermolecular forces exerted by one molecule on its neighbors. When such a solid is placed in contact with a solvent in which it dissolves, molecules at the surfaces of crystals in contact with the solvent can escape the potential well of the crystal if they possess sufficient kinetic energy, and can enter the potential well of the solvent. The rate of this escape from the solid phase is proportional to the surface area of the solid and to the "concentration" of substance in the molecular solid, which is constant.

Thus,

Ratesolid → soln = ksolid → soln*SA

Similarly, dissolved molecules collide with the surface of the solid and return to the solid phase at a rate given in 10-8-10.

Ratesoln → solid = ksoln → solid*SA*[substance]

At equilibrium, when the two opposing rates are equal,

Keq = ksolid → soln/ksoln → solid = [substance]

Thus the equilibrium constant is simply the concentration of substance dissolved in the solvent.

As we might expect from our work with phase diagrams, a change in pressure has little effect on the equilibrium established between a pure solid and a saturated liquid solution of the solid.

$$B(s) \rightarrow B(aq)$$

Both the solid and solution (liquid) phases are condensed, hence incompressible—their volumes are essentially unaffected by pressure. Unless there is a change in volume with pressure, there can be no change in concentration (the number of molecules per unit volume); and unless there is a concentration change in a phase, there can be no change in the rate at which molecules or ions leave the phase as long as T is constant.

All phase equilibria are dynamic equilibria involving a balance of opposing rates. Unless one of the rates is affected by pressure, a change in P will have no effect on the position of equilibrium. Since neither the forward nor the reverse rate of an equilibrium involving two condensed phases is affected by pressure, this type of equilibrium is unperturbed when P is varied.

The equilibrium between a solid and its saturated solution is of this type, where in this case the two balanced rates are the rate of dissolution and the rate of crystallization. The equilibrium between s and l phases of a pure substance is very similar. The temperature at which such an equilibrium exists is unaffected by even huge changes in P. Thus the solubility of a solid in a liquid should be independent of pressure.

Solutions of Gases in Liquids

The solubility of a gas in a liquid is markedly affected by the pressure at which the gas is maintained over the liquid. This is a consequence of the effect of pressure on the volume, hence the concentration, of a gas. We begin with a quantity of water at 25°C in a cylinder fitted with a piston and a gas inlet. Initially, the space above the water is free of all gas except water vapour, which exerts the characteristic vapour pressure at 25°C. We now fill the space above the water with a gas, B(g), to some pressure PB1 by opening the gas inlet valve, while holding the piston fixed in place, until the desired pressure is reached. As soon as the gas is admitted to the cylinder, gas molecules begin to strike the walls of the cylinder and the surface of the water with a frequency that depends on the concentration (pressure) of gas molecules in the space above the liquid. Some of the molecules that strike the water surface enter the liquid—i.e., they dissolve. The rate of dissolution is directly proportional to the pressure of the gas and to the surface area, SA, of the water, as indicated in equation.

Ratedissolution = k_1*SA*PB1

Here k1 is a proportionality constant (a rate constant). Gas molecules that dissolve move randomly in the liquid phase, and at any given time, some of them will be moving toward the surface. Those with sufficient energy will escape the liquid and reenter the gas phase. The rate of escape is directly proportional to the surface area of the liquid, and to the number of molecules travelling toward the surface, which in turn is directly related to the concentration of dissolved gas.

Thus,

Rateescape = k_2*SA*[B]

where k2 is also a rate constant. Immediately after the gas is introduced to the cylinder, the rate of escape is 0 because there is no dissolved gas, but the rate of dissolution is greater than 0 because P > 0. The initial imbalance of rates leads to a buildup of dissolved molecules, an increase in [B], and a corresponding increase in Rateescape, until Ratedissolution = Rateescape. At this point, dynamic equilibrium between the gas and solution phases has been established. The dynamic equilibrium between a gas and its solution is expressed.

B(g)⇔ B(soln)

where the double arrow shows the opposing equal rates. Equating the right sides and rearranging gives

$[B]/P_{B1} = k_1/k_2 = K$

Equilibrium is characterized by a particular ratio of the concentration of dissolved B to the pressure of B above the solution.This doubles the pressure of the gas above the solution, and immediately doubles Rdissolution because the number of collisions of gas molecules with the surface is now twice as great. However, Rescape is initially unaffected. The immediate effect of the pressure increase is thus to make Rdissolution > Rescape. This imbalance results in a net flow of molecules from the gas to the liquid phase, causing a gradual buildup in [B], until the two opposing rates again become equal. The new equilibrium is established with higher values of both P and [B], but their ratio, given by k_1/k_2, is the same as before the pressure increase. Thus the concentration of dissolved gas—its solubility—is directly proportional to the pressure of gas above the solution. We express this in mathematical terms by writing equation somewhat differently.

$$[B] = KP_B$$

The verbal statement that it summarizes, is known as Henry's Law. K is called the Henry's Law constant. The values of K for dissolution of several gases in water. For a gas pressure of 1 atm, these constants are numerically the same as the molar solubilities of the gases.

Table. Henrys Law Constants (Solubilities) of Several Gases in Water at 20 oC, with 1 atm Gas Pressure

Gas	K(M/atm), × 103
N_2	0.69
CO	1.04
O_2	1.38
Ar	1.50
Kr	2.79

Solutions of Ionic Solids in Liquids; The Solubility Product, Ksp. The Solubility Rules. Almost all salts are strong electrolytes; they dissolve as separate positive and negative ions. However, it is easy to show experimentally that although some salts are very soluble in water, others dissolve to only a small extent. For example 34.7 g of sodium chloride, NaCl, dissolves in 100 mL of water at 10°C, compared with only 0.000089 g of silver chloride, AgCl. Chemists have developed empirical rules that can be used to decide whether or not a salt is soluble. These are presented below. For application of these rules, we consider a salt to be soluble if its solubility exceeds 0.01 moles per liter of solution. Salts that dissolve to a lesser extent than 0.01 M are insoluble.

Solubility rules,

A salt is soluble if it contains:

A group 1 metal

NO_3^-, ClO_3^-, ClO_4^-, or CH_3COO^- (acetate).

Exceptions are $AgCH_3COO$ and $KClO_4$.

Cl^-, Br, or I^-. Exceptions for Ag^+, Pb^{2+}, Hg_2^{2+}.

SO_4^{2-}. Exceptions are the group 2 metals, Ag^+, and Pb^{2+}.

A salt is insoluble if it contains:

O_2^- or OH^-. Exceptions are the group 1 and 2 cations.

CO_3^{2-}, PO_4^{3-}, S_2^-, and SO_3^{2-}. Exceptions are the group 1 cations and NH_4^+.

The solubility rules, based on empirical observation, are useful for predicting the outcome of mixing solutions of electrolytes. This is applied in chemical analysis, in which selective precipitation of an ion from solution can enable us to determine the amount of the ion that was initially present by weighing the precipitate.

Although many ionic salts show substantial solubility in water, a number of simple and common salts fail to dissolve. These are called slightly soluble salts. Combinations of OH^-, CO_3^{2-}, and S_2^- with many cations, particularly those of the transition metals, are examples of such salts. We now want to examine the quantitative aspects of their dissolution, using the principles of equilibrium. Consider adding a quantity of the salt MX(s) (Mp+ is the cation, Xp- is the anion) to a beaker of water. Assume that we know how much of the salt will dissolve in the water, and that the amount added is more than this. What happens at the molecular level after the salt is placed in contact with the water? Upon contact with water, ions on the surface of the crystals begin to escape into the liquid at a rate that depends upon the forces of attraction between these ions and other ions in the crystal on one hand, and the ions and water molecules on the other. The rate of dissolution is proportional to the surface area, SA, of the solid salt, and the concentration of salt, which is constant in the solid.

$$Rate_{dissolution} = k_{dissolution} * SA$$

The rate of dissolution remains essentially constant as long as solid is present and the temperature is maintained constant. As dissolution proceeds, the concentrations of the ions in the liquid phase gradually build up. These ions move randomly in the liquid phase, occasionally colliding with the surface of the solid, where some of them stick—i.e., they crystallize. As their number builds, so does the rate at which they collide with the surface of the solid and crystallize.

The rate of crystallization is proportional to the exposed surface area of solid, and to the product of the concentrations of positive and negative ions. The product is required because positive and negative ions must collide at the surface in order for crystallization to occur.

$Rate_{crystallization} = k_{crystallization}*SA*[M^{p+}][X^{p-}]$

A situation in which dissolution and crystallization occur with unequal rates. Eventually, rates of dissolution and crystallization become equal, and there is no further net dissolution of solid. At this point dynamic equilibrium has been established between the solid and its ions in solution. The equilibrium is represented in equation, where the double arrow shows the two opposing rates.

$MX(s) \Leftrightarrow M^{p+}(aq) + X^{p-}(aq)$

Graphically the equalization of the rates of dissolution and precipitation over time. The equilibrium in 10-8-20 is certainly a phase equilibrium, because it involves two phases: a solid phase, represented by the salt, and a liquid phase, the solution. Equating the expressions for the rates of dissolution and crystallization, the expression for the equilibrium constant is obtained:

$K_{eq} = K_{sp}$ = kdissolution/kcrystallization = $[M^{p+}][X^{p-}]$

We give a special symbol to the equilibrium constant for dissolution of a slightly soluble salt: K_{sp}. The subscript, sp, stands for "solubility product' because it is the PRODUCT of ion concentrations, which are related to the molar SOLUBILITY of the salt. The molar solubility is the number of moles of salt that dissolve per liter of solution. The solubility product expression for the slightly soluble salt, bismuth sulfide, which dissolves according to,

$Bi_2S_3(s) \Leftrightarrow 2Bi^{3+} + 3S^{2-}$,

is Ksp = $[Bi^{3+}]$2$[S^{2-}]$3. The value of Ksp is 2×10^{-70} at 298 K, indicating that the concentrations of ions in equilibrium with solid bismuth sulfide are very small.

Before considering calculations using solubility product

equilibria, we define a few terms used to describe solutions of salts. Consider that we have a beaker containing one liter of pure water.

To this we add a quantity of a salt, MnXm(s). Suppose first that the quantity of solid added is less than the amount that will dissolve in one liter of water. We will see a gradual disappearance of solid salt as dissolution occurs, until all of the solid is gone.

The resulting solution, containing less than the maximum possible amount of dissolved salt, is said to be unsaturated. In such a solution, the product $[M^{p+}]n[X^{q-}]m$ is less than K_{sp} for the salt. Now suppose that we add more solid to the beaker. Since the solution is unsaturated, we expect that some of the added solid will dissolve, until the solution contains the maximum possible amount.

This amount depends on the solubility of the salt in water at the water temperature. We can identify this situation visually, because we will see a quantity of undissolved solid on the bottom of the beaker that does not change in size with time. This indicates that there is no longer any net dissolution of solid. At the molecular level, of course, a dynamic conversion between solid salt and dissolved ions is taking place, in which salt is dissolving and ions are combining to form solid at equal rates.

At the macroscopic level, as is true for any equilibrium, the dynamic aspects of the microscopic balance are not apparent, and it appears as if nothing is happening. A solution containing the maximum possible amount of dissolved solid at a particular temperature is said to be saturated. In a saturated solution, it must be true that the product of ion concentrations, each concentration raised to the appropriate power, is equal to K_{sp}. Finally, suppose that we are able somehow to cause more than the maximum possible amount of salt to dissolve.

In such a solution, which is said to be supersaturated, the product of ion concentrations exceeds K_{sp}, and the system is not at equilibrium. Eventually, we expect precipitation (solid formation) to occur until the product of ion concentrations becomes equal to K_{sp}, at which point equilibrium will have been established.

How do we prepare a supersaturated solution of a salt? Suppose that the salt becomes more soluble in water as the temperature of the water increases. Both sodium acetate, $NaC_2H_3O_2$, and alum, $KAl(SO_4)_2$, qualify, as do many other salts. We consider sodium acetate. The solubility of sodium acetate in water at 25 °C is 125 g per 100 mL of water. At 100 °C (the boiling point of water) the solubility increases to 170 g per 100 mL of water. If we weigh an amount of sodium acetate that is greater than 125 g but less than 170 g, add it to 100 mL of water, and warm the mixture with stirring to near 100 °C, all of the sodium acetate should dissolve. (In order to observe supersaturation, it is necessary to remove all solid material from the solution while it is hot.

This can be done by careful filtration of the hot solution.) We now carefully cover the solution to exclude dust and dirt, and allow it to slowly cool to room temperature. With luck, all of the sodium acetate stays dissolved, even though the equilibrium solubility at 25 °C is exceeded. This is a supersaturated solution. How can a substance remain dissolved when it should precipitate? Why does the salt not crystallize from solution as the system cools, so as to always maintain the solubility product equilibrium? The answer is interesting—and simple. Formation of crystals of solid from ions or molecules dissolved in a solvent is similar to the freezing process, in which solid forms from liquid:

$$l \Leftrightarrow s$$

For this process, $\Delta PE < 0$, favoring formation of solid; but $\Delta S < 0$ (because solid is more ordered than liquid), opposing formation of solid. Supersaturation is thus analogous to supercooling of liquids. For crystallization of a salt to begin, it is necessary that positive and negative ions, moving randomly in solution, coalesce in the proper arrangement for a crystal to begin to form. Because this process is opposed by entropy, it often takes some time, during which the solution can hold more dissolved material than it should. Once the beginnings of a crystal form, however, this serves as a nucleus, or template, on which other ions can deposit, and crystallization proceeds rapidly.

Thus we can initiate crystallization of sodium acetate from the supersaturated solution by adding a small crystal of sodium acetate as a template. Crystallization then proceeds until the Ksp is satisfied (i.e., until the solution is saturated). (Often, addition of any small solid particles to a supersaturated solution promotes crystallization.

That is why filtration to remove dust is necessary.) Generally, the more complicated and/or unsymmetrical the structures of the ions of the salt, the more difficult it is for the ions to coalesce in the correct way to initiate crystallization. Thus we expect salts of complex ions to more readily form supersaturated solutions than salts of simple ions. In agreement with this guideline, it is very difficult to prepare a supersaturated solution of NaCl; whereas preparation of supersaturated sodium acetate, which contains an anion of complex structure, is readily accomplished.

Chapter 5

Properties of Solids

OPTICAL PROPERTIES

A particle placed in a beam of light will scatter some of the light incident on it (i.e. the light will change direction) and absorb some of the light (i.e. the electromagnetic energy is transformed into other forms of energy by the particle). The light we then receive has been attenuated or, to put it another way, suffered extinction. The amount of light scattered and/or absorbed by the particle depends, in an intricately involved way, on the exact nature of the particle, and also the nature of the incident light. Consequently the extinction is dependent on the chemical composition of the particle and its size, shape and orientation, together with the wavelength and polarization state of the light. Therefore, we must include all these factors in our interpretations of spectra.

Identification of astronomical dust is usually achieved by comparing the observed spectral data with a catalogue of laboratory spectra of materials considered to be similar to those found in the astronomical dust. A computer programme can be used to make correlations between the reference spectra and the observed spectra, however there are limitations to this technique.

In order to compare the laboratory and observed spectra they need to be of similar type, i.e. transmission spectra of very fine particulate materials. Many of the available spectra are obtained from samples for which no attempt has been made to simulate the possible conditions in the observed astronomical dust. One of the reasons for this is that the laboratories want to optimize the surface conditions of the sample to yield the maximum amount of spectral information. Also, the applications to which infrared spectra are put are diverse and therefore the collection of such spectra uses many different techniques, producing different types of spectra, according to the application. A complete understanding of the transmission and reflection properties of both particulate and bulk media is imperative for the correct interpretation of the spectral data.

PHYSICAL OPTICS

Absorption, reflection, scattering, polarization and dispersion of light by a material medium can all be explained in terms of the atoms, molecules and lattice structures which make up the medium. This section will concentrate on the mechanisms by which electromagnetic radiation is absorbed by matter.Absorption of radiation by matter always involves the loss of energy by the radiation and a gain in energy by the atoms or molecules of the medium.

The energy of an assembly of atoms consists partly of thermal energy and partly of internal energy associated with the binding of the extra-nuclear electrons to the nucleus and with the binding of the particles within the nucleus itself. Molecules have, in addition, energy associated with the oscillations of the atoms within the molecule with respect to one another, and solids have energy associated with intermolecular bonds. The energy absorbed from radiation appears as increased internal energy or as increased vibrational energy of interatomic or intermolecular bonds.

Classical Electromagnetic Theory

In classical electromagnetic theory, atoms and molecules are considered to contain electrical charges (i.e. electrons, ions) which are regarded as oscillating about positions of equilibrium, each with its appropriate natural frequency, ν_o . When placed in a radiation field of frequency , each oscillator in the atom or molecule is set into forced vibration with the same frequency as that of the radiation.

If $\nu << \nu_0$ or $\nu >> \nu_0$, the amplitude of the forced vibration is small, but as ν approaches ν_0, the amplitude of the forced vibration increases rapidly.

To account for the absorption of energy from the radiation field, it is necessary to assume that the oscillator in the atom or molecule must overcome some frictional force proportional to its velocity during its forced motion. For small amplitudes of forced oscillation, when ν is very different from ν_0, the frictional force, and therefore the absorption of energy, is negligible. Near resonance (i.e ν is approx. equal to ν_0), the amplitude of oscillation becomes large, with a correspondingly large absorption of energy to overcome the frictional force. Therefore, the radiation of frequencies near the natural frequency of the oscillator corresponds to an absorption band.

Acoustic and Optical Modes

Now, let us examine the allowed vibration modes in a lattice. The most rudimentary lattice we can consider is a linear chain of identical atoms.

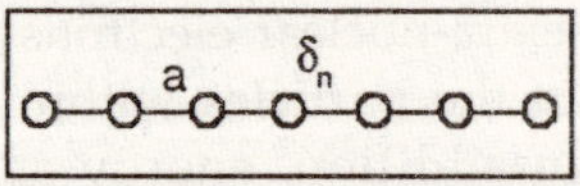

Fig. A simple chain lattice of identical atoms

If we displace the nth atom from its equilibrium position by a small amount ðn parallel to the chain, the neighbouring atoms ((n-1)th and (n+1)th) will also be displaced by $\delta_n - \delta_{n-1}$ and $\delta_n - \delta_{n+1}$ respectively. Therefore the nth atom is actually displaced by an amount $\delta_n - \delta_{n-1}$ with respect to the (n-1th) atom, and by $\delta_n - \delta_{n+1}$ with respect to the $(\delta + 1)^{th}$ atom. According to the Lorentz model, these atoms can be considered to be connected by springs. Therefore the n th atom will experience a restoring force F equivalent to that of Hooke's Law. Because the n th atom is connected to two springs (one to each neighbouring atom) there will be two contributions to this force, which is given by:

$$F = -\eta\, [(\delta_n - \delta_{n-1}) - (\delta_{n+1} - \delta_n)]$$

where η is the spring constant. Applying Newton's second law we also get:

$$F = m \cdot d^2ð_n/dt^2$$

where m is the mass of the oscillator.

Thus,

$$d^2ðn/dt^2 = (\eta/m)\,[(ð_n - ð_{n-1}) - (ð_{n+1} - ð_n)]$$

Solutions to equation have the form:

$$ð_n = ð_o\; e^{i(kna - \omega t)}$$

where is the angular frequency of the wave, k is the wavenumber and ðo is the amplitude of oscillation. If we now substitute equation we get:

$$\omega^2 = 2(\eta/m)\,[1\text{-}\cos(ka)]$$
$$\omega^2 = 4(\eta/m)\,\sin^2(ka/2)$$
$$\omega = \pm 2\,(\eta/m)^{1/2}\sin(ka/2)$$

This equation gives the allowed vibration modes in this simple lattice. From equationit is obvious that $\omega > 2(\eta/m)½$ is not possible. Frequencies above $\omega = 2(\eta/m)½$ cannot propagate through the lattice. In this model the displacement of the oscillator is along the chain, therefore the wave motion is longitudinal. This is the longitudinal acoustic mode.

Now consider a chain in which there are two different atoms of type X and Y, with masses m_X and m_Y respectively ($m_X > m_Y$), placed alternately along the chain. Again we displace the nth atom, of type X, by a small amount $ð_n$. The nearest neighbouring atoms, both of type Y, are displaced by $ð_{n-1}$ and $ð_{n+1}$ as before; the next nearest atoms, both of type X, are displaced by $ð_{n-2}$ and $ð_{n+2}$ and so on. We have two different atom types in the chain, so we require two equations for the restoring force: one for the nth atom of type X and one for the $(n+1)^{th}$ atom of type Y. Hence we get:

$$F_X = -\eta_X[(ð_n - ð_{n-1}) - (ð_{n+1} - ð_n)]$$
$$F_Y = -\eta_Y[(ð_n - ð_{n+1}) - (ð_{n+1} - ð_{n+2})]$$

For simplicity, let $\eta_X = \eta_Y = \eta$. Again we can use Newton's second law and get:

$$F_X = m_X \cdot d^2ð_n/dt^2$$
$$F_Y = m_Y \cdot d^2ð_{n+1}/dt_2$$

And again, analogous to the monatomic chain we get solutions of the form:

$$\eth_n = \eth_X \, e^{i(kna - \omega t)}$$
$$\eth_n = \eth_Y \, e^{i(k(n+1)a - \omega t)}$$

From which we get:

$$\omega^2 = \eta[(1/m_X) + (1/m_Y)] \pm \eta[\{(1/m_X) + (1/m_Y)\}^2 - \{4\sin^2(ka)\}/m_X m_Y]$$

If we take the negative sign in equation, we get the acoustic mode already encountered, with the maximum allowed frequency at $= 2(\eta/m_X)^{1/2}$ again. However, if we take the positive sign in equation, we get the optical mode, so called because this mode is easily excited by radiation of the appropriate frequency in ionic solids. In this case we find that there is a range of permitted frequencies with both a maximum and a minimum which correspond to the longitudinal and transverse optical (LO and TO) modes.

The minimum frequency (TO) is given by:

$$\omega TO = 2(\eta/m_Y)^{1/2}$$

The maximum frequency (LO) is given by:

$$\omega LO = 2[\eta/\{(1/m_X)+(1/m_Y)\}]^{1/2}$$

The transverse and longitudinal optical modes represent the onset and cutoff of an absorption band.

It is generally thought that absorptivity measurements only involve the TO component, however, in reality the LO component can play an important role. The contribution from the LO mode depends on the constituents of the molecule/solid under consideration as well as the crystal structure, grain size/sample thickness etc. The conditions under which the spectrum is taken and the nature of the spectrometer also affect the relative contributions from the TO and LO modes. It is therefore necessary to understand the contributions from all these factors in order to interpret spectra correctly.

ELECTRICAL PROPERTIES OF MATTER

At the beginning of this section the mechanism by which electromagnetic radiation loses energy to material media was outlined. We have just concluded a simple discussion of the permitted frequencies in a material medium. Now we need to combine these two: we have a system in which atoms or molecules are connected by "springs" and that are forced to oscillate in an electric field. The motion is affected by friction in the guise of a damping force. If the oscillator has mass m and charge q, we can expect an equation of motion of the form:

$$m\,(d^2ð/dt^2) + b\,(d\,ð/dt) + \eta ð = qE$$

where ð is the displacement of the oscillator, b is the damping force, η is the spring constant and E electric field of the electromagnetic radiation incident on the material. The electric field is assumed to be time harmonic with frequency ω. From equation we can see that η is approximately equal to $m\omega_o^2$ (where ω_o is the natural frequency of the medium); the damping force can be written as b = γm, where γ is the damping coefficient.

Thus, equation becomes:

$$d^2ð/dt^2 + \gamma(d\,ð/dt) + \omega_o^{\,2}ð = qE/m$$

The solution of equation is:

$$ð = (-qE/m)\,.\,[1 / \{(\omega_o^{\,2} - \omega^2) - i\gamma\,\omega\}]$$

which gives us an induced dipole moment, p, of:

$$p = (-q^2E/m).[1/\{(\omega_o^{\,2} - \omega^2) - i\gamma\,\omega\}]$$

If we assume that the displacement ð is small we can also assume that there is a linear relationship between the induced dipole moment and the electric field, which gives:

$$p = a(\omega)E$$

where a(ω) is the frequency dependent polarizability of the atom or molecule. Combining equations:

$$a = (-q^2/m)\,.\,[1/\{(\omega_o^{\,2} - \omega^2) - i\gamma\,\omega\}]$$

Because of the inclusion of the damping term, the polarizability is a complex quantity. If there are N oscillators per unit volume, the macroscopic polarizability, P, is given by:

$$P = N\langle p\rangle = N\,a\,E$$

From Maxwell's equations we can get a complex displacement, D, given by:

$$D = E = E + 4P$$

where is the complex dielectric function. Substituting for P from equation we get:

$$E = E + 4N\,a\,E$$

$$= 1 + 4N\,a$$

now substituting from equation we get:

$= 1 + (4N\ q^2)/m.[1/\{(\omega_o^2 - \omega^2) - i\gamma\ \omega\}]$

The dielectric function is complex and can, therefore, be written as:

$= \in' + \in''$

From equation we get:

$\in' = 1 + (4N\ q^2)/m.[(\omega_o^2 - \omega^2)/\{(\omega_o^2 - \omega^2)^2 + \gamma^2\ \omega^2\}]$

$\in'' = 1 + (4N\ q^2)/m.[(\gamma\omega)/\{(\omega_o^2 - \omega^2)^2 + \gamma^2\ \omega^2\}]$

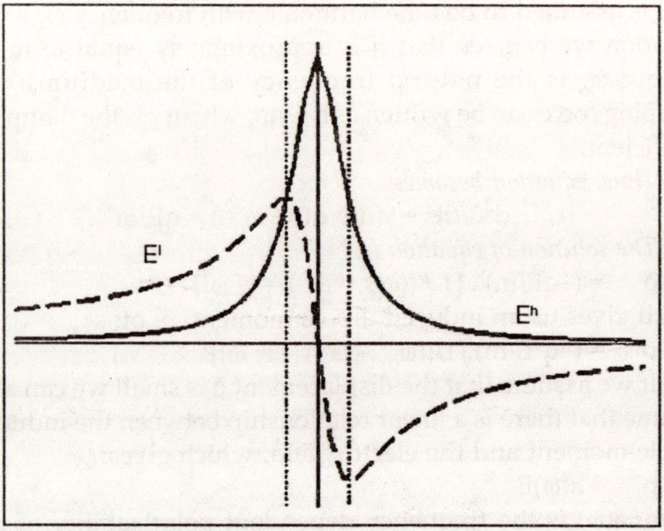

Fig. The variation of the real and imaginary parts of the dielectric function($\in'$ and $\in''$) with respect to frequency ω.

We have expressions for the real and imaginary parts of the complex dielectric function $\in'$ and $\in''$. What do these expressions tell us about the response of a material to electromagnetic radiation? Let us consider the frequency dependence of $\in'$ and $\in''$.

The frequency dependences given by equations which shows that, for most of the spectrum, increases with increasing frequency ω. This is known as normal dispersion. However, for a small section close to the resonance frequency of the oscillator, ω_o, the value of $\in'$ decreases with increasing ω.

This is known as anomalous dispersion. The width of the region of anomalous dispersion is determined by the strength of the damping coefficient, γ. For small values of γ, the region of anomalous dispersion is narrow.

However, for large values of , the anomalous dispersion covers a broader range of frequencies, and hence the region over which absorption by the medium extends is broader. Inspection of equations also shows us that, for small values of γ, the value of ϵ'' remains small and the peak of ϵ'' is very close to the resonance frequency of the medium, ω_0. Likewise, has its anomalous region centred at about ω_0. However, for large values of γ , the situation is quite different.

With increasing γ, the height of the peak of ϵ'' becomes much larger and the peak position moves away from the resonance frequency, as does the position of the anomalous region in ϵ'. Sinceγ is the measure of the damping within a medium, i.e. the mechanism by which the medium absorbs energy from electromagnetic radiation, large values of γ correspond to highly absorbing media.

Similarly, low values of correspond to relatively transparent media. Therefore, for relatively non-absorbing materials, a region of anomalous dispersion, and therefore an absorption feature, coincides approximately with the resonance frequency of the oscillators.

Furthermore, for highly opaque, absorbing materials, the anomalous dispersion and the absorption feature will be shifted away from the resonance frequency of the oscillators.

As can easily be seen from equation the physical property of a solid that determines its dielectric function is its electric polarizability, which, in turn, consists of contributions from the individual dipole moments of the oscillators within the solid. It is therefore useful to consider what actually constitutes these dipole moments. There are three possible sources of individual dipole moments in solids to examine:

- The dipole moments due to the orientation of polar molecules in the solid;
- Ionic polarisation due to the relative movement of the ions in the solid;
- Electronic polarisation due to the displacement of atomic electrons relative to the nuclei. Depending on

the nature of the solid, the first two considerations may not be applicable (i.e. the solid may not contain polar molecules or ions), however all solids contain electrons and nuclei and therefore electronic polarisation is found in all solids. The contributions to the dielectric function from these three types of "oscillators" lead to features in three separate regions of the spectrum. The dipole moments of polar molecules are expected to give rise to features in the $\sim 10^{11}$Hz (~3mm) region of the spectrum; likewise the ionic polarisation affects the spectrum in the $\sim 10^{13}$Hz (30μm) region (the infrared); and the electronic polarisation will make contributions at $\sim 10^{15}$ (0.3μm; UV). Therefore we expect some solids to exhibit spectral features in all these regions.

REFRACTIVE INDICES OF MATERIAL

There are two sets of optical constants that are closely interrelated. These are:

- The real and imaginary parts of the complex refractive index n = n - ik;
- The real and imaginary parts of the complex dielectric function (relative permittivity) = +.

They are related by:

∈′ = n2 – k2 = Re [n2]

∈″ = 2nk = Im [n2]

Which of these two sets of constants are most applicable depends on the material under investigation. Reflection and transmission by bulk media are best described using the complex refractive index (n – ik), whereas absorption and scattering by particles which are small compared with the wavelength are best described by the complex dielectric function = ∈′ + ∈″, however, they remain interchangeable. The media with which we are concerned have complex refractive indices that depend on wavelength and are complex.

We can write the refractive index, n , of a medium as:

n = n – ik

To simplify the problem we assume that we have radiation

incident on our medium in the form of an idealised, sinusoidal wave.

The incident radiation can be written as:

A = Aoe i(ωt – 2x)

where A is the amplitude, is the angular frequency and the wave is propagating in the positive x-direction.

Once the radiation enters the medium, the velocity of light becomes c/ n, so that:

A = Aoe i(ωt - 2n x)

A = Aoe iωt – 2i(n – ik) x)

A = Aoe i(ωt – 2n x /) e-2 kx

From this we can see that the intensity of the wave (which is proportional to A2) will decrease exponentially with decay constant of 4k. Therefore the imaginary part of the refractive index, k, is directly responsible for the absorption of the light in the medium.

SCATTERING

Another consideration in the spectra of small particles is scattering. Consider a small particle placed in a beam of light. The intensity of the elctromagnetic radiation incident on the particle is Io; at a point a large distance r from the particle, the intensity of the scattered light is I. The scattered intensity I must be proportional to the incident intensity Io, and the inverse square of the distance 1/ r 2, hence we can say that:

I = [Io *F*(θ, ϕ]/[k 2 r 2]

where k = 2 is the wavenumber, and *F*(θ, ϕ) is an orientation dependent, dimensionless function of the direction but not distance from the particle.

The total energy scattered by the particle in all directions is defined as being equal to the energy of the incident light falling on the cross-sectional area Csca, where Csca is the scattering cross section.

If the function *F*(θ, ϕ) is divided by k2Csca the phase function is obtained, which has no dimensions and whose integral over all directions is 1.

Therefore:

Csca = (1/k2).§ *F*(θ, ϕ) dω

where $d\omega = \sin\theta \, d\theta d\phi$ is the element of solid angle and the integral is taken over all directions.

In a similar way to the scattering cross-section case, the energy absorbed by the particle can be put equal to the energy incident on the area Cabs (the absorption cross-section). Likewise the total energy depleted from the original beam can be put equal to the incident energy on the area Cext, the extinction cross-section.

The law of conservation of energy requires that:

Cext = Cabs + Csca

The efficiency of a particle to absorb, scatter and extinguish are given by the Q factors, where:

Cext = Qext x area

Cabs = Qabs x area

Csca = Qsca x area

Therefore:

Qext = Qabs + Qsca.

The Q-factors can be calculated from the complex refractive index (or complex dielectric function) using Mie Theory

MIE THEORY

Mie theory describes the way in which spherical, homogeneous particles interact with electromagnetic radiation. Certain assumptions are made about these particles in order to simplify the situation.

These assumptions are:

- Only the interactions of a single particle with light of arbitrary wavelength are considered.
- The optical properties of the particle are completely described by frequency-dependent optical constants (i.e. the complex refractive index or the complex dielectric function).
- Scattering is elastic, i.e. the frequency of the scattered light is identical to that of the incident light.
- Only single-scattering occurs. That is, each individual particle is acted on by an external field (due to the

incident radiation) in isolation from the other particles. The total scattered field is merely the sum of the fields scattered by each particle (i.e. the particles do not affect each other).

- For a collection of particles, the number of particles is large and their separations are random, so that the waves scattered by the individual particles have no systematic phase relation.
- The medium in which the particles are embedded is considered to be:
  - Linear,
  - Homogeneous;
  - Isotropic.

The theory used to find the transmission spectra of small particles from the optical constants is known as Mie Theory. This calculates the Q-factors, which are the efficiency factors for extinction, scattering and absorption.

There is no simple explicit formula for the Q-factors, however if 2a, (where a is the radius of a spherical particle and is the wavelength of the incident light), then for a spherical particle:

Qsca = (8/3)(2a/)4 Re[{(n2 – 1)/(n2 + 2)}2]
Qabs = 8a/ Im[(n2 – 1)/(n2 + 2)]

where, n = n – ik.

Given the close relation between the complex refractive index and the complex dielectric function, we can also write these equations in terms of the dielectric function:

Qsca= (8/3)(2a/)4 Re[{(epsilon - 1)/(epsilon + 2)}2]
Qabs= 8a/ Im[(epsilon - 1)/(epsilon + 2)]

where epsilon = – i.

There are computer programs that use Mie Theory to evaluate the Q-factors and thus produce the relevant spectra.

These programs require the input of:

- A dimensionless length 2 8a;
- The real and imaginary parts of the refractive index of the materials n = n – ik, or the real and imaginary parts of the dielectric function of the material epsilon = ∈′ – ∈″.

DETERMINATION OF OPTICAL CONSTANTS

It is assumed that the optical properties of a particle are completely specified by frequency-dependent optical constants. The optical constants of minerals of astronomical significance must, therefore, be determined. This is most easily done using bulk materials, as opposed to particles, but either can be used. The data accumulated from published sources include both bulk and particulate samples and transmission, absorption and reflection spectra. It is not very useful to compare spectra measured by different techniques, especially since the appearance of the spectra depends on the measurement conditions.

The transmission spectrum of a bulk medium is often very different from its particulate counterpart. The chemical composition remains the same, but the state of aggregation, and hence scattering properties, differ. The homogeneous bulk does not scatter light, however the particulate material has a transmission spectrum that may be primarily the result of scattering. Gross optical properties of a given material differ appreciably depending on its state of aggregation.

Optical theory was therefore used to normalize the data by finding the intrinsic optical constants and calculating the spectra. Since most of the astronomical spectra will be particulate, either emissive or absorptive in nature, the data can be used to calculate the spectra of such systems. There are various methods, each of which has its advantages and limitations, depending on the form of the material we are interested in, are outlined here:

- Measurement of refractive angles. The real part of the refractive index, n, can be obtained using Snell's Law. This requires the sample to be of high transparency (i.e. $k \sim 0$).
- Measurement of transmittance and reflectance of a slab for light at near-normal incidence. Again, the sample needs to be transparent in order to have a measurable transmission, but the transparency required is not as high as method 1.

- Measurement of reflectance at near-normal incidence over a wide range of frequencies. This involves using a Kramers-Kronig relation to obtain the phase shift of the reflected light. The sample must be highly opaque and the method requires measurements over an extended wavelength region and usually extrapolation into unmeasured regions.
- Ellipsometric techniques. These measure the amplitude ratios and phase shifts directly. They are difficult to use over large wavelength regions due to instrument limitations.
- Measurements of reflectance for incident light of various polarisation states and two oblique angles of incidence. This method generally requires large sample surfaces, although the use of this method with small surfaces has been achieve by utilising cryogenic detectors.

Many spectra have been obtained for use with astronomical observations and are therefore already the absorption/emission spectra for particulate material. However, the exact nature of the spectrum must be taken into account. While they already represent spectra of particulate material, the spectrum plot may be of percentage transmission, percentage absorption or absorption/extinction efficiency. Astronomical spectra are of the latter type, and therefore it is necessary to normalise all other spectra to this type. So, the question is, how do we generate the extinction efficiency, Qext, from the percentage absorption or transmission? Since the absorption is the inverse of the transmission, the method is essentially the same for either type of spectrum.

Suppose we have a ray which passes through an infinitesimally thin region of thickness dx which causes extinction. Let the intensity of the ray entering the extinguishing region be Iv. The radiation will lose intensity through interaction between the radiation and the matter in the extinguishing region, so that the intensity of radiation emerging from the extinguishing region is Iv+dIv.

The extinction coefficient, K is defined by:

$$dIv = -K v Iv dx$$

We can also define the optical depth, τ, as:

$$\tau v = \text{pathlength } Kvdx$$

The attenuation of radiation by the effects of extinction is given by e – τv . This is essentially the absorption of the radiation. Therefore:

$$A = e - \tau v$$

and lnA = –τv where A is the absorption. Since transmission is the inverse of absorption, T = 1/A, we also have:

$$\ln T = \tau v$$

So the optical depth, τ, can be obtained from the absorption or transmission. This must then be related to the extinction efficiency, Qext

The optical depth has been defined as:

$$\tau v = \text{pathlength } Kvdx$$

For a homogeneous extinguishing medium this gives:

$$\tau v = D\, Kv$$

where D is the path-length, and Kv is defined as:

$$Kv = n\, Cv$$

where n is the number of particles per unit volume and Cv is the extinction cross-section for each particle.

$$Cv = Qext\ a2$$

Hence,

$$\tau v = D\, nQext\ a2$$

and, τv is proportional to Qext. Therefore the shape of the extinction efficiency spectrum is identical to the shape of the optical depth spectrum and, lnT is proportional to Qext. From the transmission (or absorption) spectra we can find the shape and positions of spectral feature, however this is not a quantitative method unless the pathlength is known and the medium is homogenous.

LABORATORY SPECTRUM

A problem has arisen regarding the laboratory spectra of some particulate materials which seems to have come about as a result of a misunderstanding of the optical properties of solids. The usual method for obtaining spectra from a particulate sample is to disperse the sample in a pellet of a substance which is transparent to light in the wavelength range of interest. This is typically potassium bromide (KBr) for mid-infrared measurements. In 1978, Dorschner undertook a study comparing the spectra of silicate samples produced using the KBr-dispersion method with spectra produced by extracting the optical constants from bulk samples. This resulted in a mismatch between the spectra, which they took to imply that the KBr embedding created a shift in the peaks of spectral features to longer wavelength. Since then the use of this correction of spectra for KBr-dispersion has been widespread, although not universal.

The most prevalent use of this correction factor has been with respect to silicon carbide (SiC) spectra. Silicon carbide was theoretically predicted to form around carbon-rich AGB stars in 1969 and it was suggested that the ~11.5 μm feature seen in the spectra of these stars may be due to silicon carbide dust. This prompted various laboratory investigations of this solid. Following the work of Dorschner used the same procedures to investigate the differences between a KBr-dispersed particulate SiC spectrum and the spectrum derived from the bulk SiC sample. They found a shift in the peak wavelength analogous to that found by Dorschner, from which they surmised that a spectrum of SiC obtained from a KBr-dispersion sample must be corrected by shifting the peak of the feature by 0.4 μm to shorter wavelengths. The result of applying this correction factor was to give their alpha-SiC sample a feature with a peak at 11.4 μm rather than the 11.8 μm peak before correction. The SiC spectra published by Borghesi were again obtained using the KBr-dispersion method. Their published spectra are not corrected for the method, but it is explicitly declared that the correction factor should be applied before the spectra could be used. Again it is found that the alpha-SiC sample has a peak position which is shifted from 11.8 μm to 11.4 μm by correction. Borghesi also presented a beta-SiC sample which has a peak position shifted from 11.4 μm to 11.0 μm by this correction. Used various existing KBr-corrected laboratory spectra to derive a complete spectrum for alpha-SiC.

The use of these various sources of laboratory spectra by astronomers led to the identification of the ~ 11.5 μm feature in carbon star spectra with alpha-SiC. However there are samples of SiC available in meteorites which are believed to have been formed around such stars, and all these grains are of the beta-SiC form. It is difficult to find a mechanism by which alpha-SiC could be transformed into beta-SiC, and thus it is assumed that the meteoritic grains were formed as beta-SiC. This constitutes a discrepancy between the astronomical and meteoritic results.

However, if the KBr-correction factor is not applied to the laboratory data, the silicon carbide type which best fits the astronomical observations switches from alpha-SiC to beta-SiC, removing the discrepancy between the astronomical and meteoritic data.

The change in the attribution of the ~ 11.5 μm feature in carbon star spectra. The effect of the correction factor has been to cause the misidentification of a dust species in space, therefore, it is necessary to understand how this correction factor arose and why it is not valid.

The original work which proposed the KBr-correction used Mie theory, which is not ideal since it assumes that the particles in the sample are spherical. It has since been shown that, for crystalline olivine, a similar method for deriving a spectrum from bulk optical constants using a continuous distribution of ellipsoids rather than spheres produces a spectrum that is almost identical to the KBr-dispersion spectrum.

Another problem with the KBr dispersion spectra of SiC arises from the high opacity of this compound. For any sample, overloading the KBr pellet with the sample causes the pellet to become opaque.

The effect of this is that the zero-transmission point is incorrectly defined. In these cases the spectrum obtained is worthless. This problem is particularly important when the sample is SiC. Silicon carbide is a very efficient absorber, therefore the amount of SiC needed to cause the KBr pellet to become opaque is quite small (i.e. less than needed for other minerals). This is compounded by particle size considerations. Due to their great affinity for absorbing electromagnetic radiation, SiC particles do not have to be very large before they appear to be opaque.

For very fine-sized particle dispersions, in reasonably low concentrations, this effect is negligible, allowing the acquisition of spectra that are apposite. A good example of the effect of overloading the KBr pellet with sample can be seen in Papoular , who presented two spectra of beta-SiC consisting of broad overlapping peaks at 10.9 and 12.2 μm, unlike any other reported spectra.

These positions are close to the TO and LO components. The very high absorbance units of 1 and 2.5 for these samples indicate overloaded pellets. For extreme concentrations of SiC (or large thicknesses), light is reflected between the TO and LO modes: the scattering in the pellet produces the dip in absorption.

Problems occur at high absorption because the partial opacity induces a frequency dependent baseline. Papoular were attempting a reassessment of the KBr correction factor, however the use of overloaded KBr pellets means that their results are erroneous.

THIN FILM VS. KBR-DISPERSION SPECTRA

The problems incurred by the use of the KBr-dispersion technique has prompted the hunt for an alternative source of applicable spectra. New infrared (IR) absorption spectra of thin films of alpha- and beta-SiC created by compression in a diamond anvil cell (DAC) will be presented here.

Unlike the KBr-dispersion method, a dispersive medium is not used. This relatively new approach is quantitative (if sufficient care is taken to produce an appropriately thin and uniform film) as shown by comparison of thin film spectra of olivines, garnets, wadsleyite, $MgAl_2O_4$ spinel, and $CaGeO_3$ perovskite to reflectivity data from the same samples.

Furthermore, measurements of thin film spectra of garnets are nearly identical to single-crystal absorption data. Moreover, spectra of thin films of AlN (aluminium nitride) are identical to the spectra of nanometre-sized particles obtained using the KBr technique.

Therefore, thin film spectra are equivalent to the collection of spectra of small particles in a vacuum, and can therefore be applied to astronomical data without further manipulation.

Laboratory Techniques

Single-crystals of alpha-SiC were purchased by A.M. Hofmeister from Alpha/Aesar. This specimen is 99.8% SiC, consisting of hexagonal platelets of 50 to 250 μm in diameter and 5 to 15 μm thick. Less than 1% of the platelets had an amber colour, the remainder were pale grey. All were transparent in the visible with smooth, highly reflective surfaces. Polycrystals of beta-SiC were donated by Superior Graphite.

The purity of this sample is also 99.8%. One batch consisted of 1 μm powder, the other was a conglomerate of equant crystallites of up to 25 μm in size. For this study, only the gray crystals of alpha-SiC were examined. Mid-IR spectra were obtained from 450 to 4000 cm^{-1} (2.5-22.2 μm) at 2 cm^{-1} (~0.01 μm) resolution using a liquid-nitrogen cooled HgCdTe detector, a KBr beam splitter and an evacuated Bomem DA 3.02 Fourier transform interferometer.

Thin films were created through compression in a diamond anvil cell which was interfaced with the spectrometer using a beam condenser. Type II diamonds were used. Film thickness was estimated from the initial grain size, by the relative relief and colour seen among the various films, and by the increase in grain diameter from the initial size during compression.

Efforts were made to cover the entire diamond tip (0.6 mm diameter) with an even layer of sample, but slight irregularities in the thickness were inevitable. Reference spectra were collected from the empty DAC. Uncertainties in peak positions are related to peak widths because the accuracy of the FTIR spectrometer is high, ±0.01cm^{-1}.

Spectra obtained from alpha-SiC have an intense, broad band near 11.8 μm. The peak position lies between the longitudinal optic mode (LO) and transverse optic mode (TO) components and a shoulder is seen at the LO position. A shoulder also occurs at 12.2 μm.

The sample thickness could not be precisely determined, but was estimated to be sub-micron. Spectra obtained from beta-SiC) depend somewhat on thickness. For the thinnest films, of sub-micron thickness a fairly symmetric peak is found at 11.3 to 11.4 μm, and a weak shoulder exists at 10.7 μm, consistent with excitation of the LO component.

Spectra from thicker film samples, ~1 μm in thickness from visual inspection , have a peak at a similar position, with an asymmetric increase in intensity on the short-wavelength side, and display additional weak features. The 12.7 μm band is due to the TO feature.

The weak, broad band at 13.4 microns is not an absorbance feature but is due to the Christiansen effect which gives a minimum when the real part of the index of refraction is unity. The asymmetry of the main peak is due to the baseline rising towards the visible, probably a scattering effect from the grain boundaries. A spectrum from the thickest sample examined ~1 μm), had high absorbance values overall, with the Si-C stretching peak superimposed.

The appearance of the peak is intermediate between the peaks observed from the thin and moderately thick samples in that the main peak is symmetric but that a weak subsidiary feature exists at 12.7 μm.

Below 8 μm, the absorbance drops, rather than increasing as in the other spectra, because of interference fringes in the near-IR. These interference fringes indicate a distance of 5 μm, inferred to be the separation of the diamond anvils. The peak positions of the beta-SiC samples are relatively independent of the thickness. No difference can be discerned between the two samples of beta-SiC. Additional spectra from both beta-samples resembled those shown.

The appearance of the spectra are consistent with being due to pure absorption for the thinnest samples and absorption with minor reflection for the thicker samples, given that the LO-TO coupling is stronger in beta-SiC than in alpha-SiC.

Comparison with Dispersed-sample Results

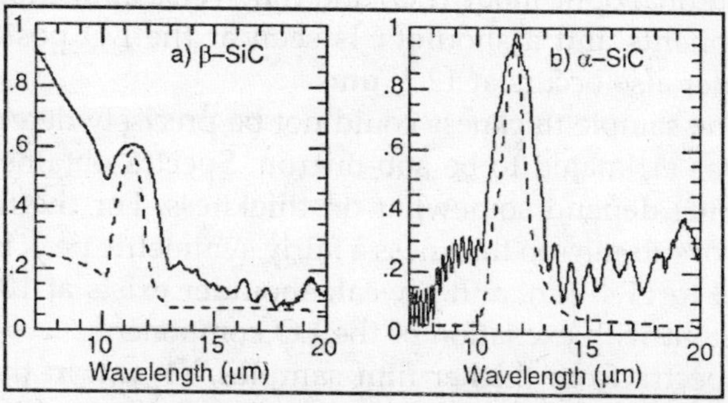

Fig. Thin film Spectra (Solid Lines) and the Uncorrected Borghesi KBr Dispersion Data

KBr matrix spectra of beta-SiC obtained by Borghesi for a fine grain size sample (mean diameter modeled by them as 0.02 μm, average diameter observed in TEM as 0.12 μm) closely match our own thin-film data, particularly the spectrum. The greatest difference is that the TO mode appears as a shoulder, rather than a separate peak. This difference is obviously due to sample thickness, because the thinner film has a barely discernable shoulder at the TO position. The LO mode occurs as a weak shoulder in their dispersion data, but this was not noted by them.

Their peak barycenter was at 11.4 μm, which equals the results from our thin films. The match of Borghesi is consistent with an estimated film thickness <0.1 μm. The beta-SiC spectrum of Papoular, that has maximum absorbance of 0.4 (i.e. the pellet is not overloaded), has a peak at 11.5 μm.

For alpha-SiC, the KBr-dispersion spectrum of Borghesi's smallest-grained (mean diameter modeled by them as 0.04 μm, average diameter observed in TEM as 0.16 μm) and purest sample (SiC-600) closely matches the spectrum of our thinnest film. Its peak position of 11.6 μm equals our result, given the experimental uncertainties. The positions of the shoulders are comparable to the LO and TO positions. Their sample N is compromised by ~10% impurities (C and SiO2). Their SiC-1200 sample was 3-10 times larger grained, even for the ground and sedimented fraction, and is inappropriate for comparison.

The study by Friedemann involved larger grain sizes, but yielded similar spectral profiles, with a slight shift of the peak position to 11.8 μm. It is clear that the application of a KBr matrix wavelength correction is incorrect, since the barycenter peak for KBr dispersions with fine grain sizes and reasonably low concentrations equals that of corresponding thin films, while the peak shapes are in excellent agreement.

For these (<0.1 μm) grain sizes or film thicknesses, bulk absorption rather than surface effects dominates in the vicinity of the intense peak. Further, Bohren and Huffman incorrectly assumed that absorption peaks occur at the maximum in the dielectric function, which is neither expected from theory nor observed experimentally. Also, the width in epsilon double prime was assumed to be zero, which is particularly inappropriate for intense IR modes as seen in SiC. The maximum in defines the TO position, whereas the bulk absorption peak is shifted towards the LO position.

The shift correlates with the strength of the mode: for SiC, the effect is pronounced, with the absorption lying roughly midway between the TO and LO modes. Only for extremely thick or large grain samples, ~1 μm, do the parameters of the dispersions differ from those of a bulk sample but the differences are due to internal scattering among the particulates and sampling effects due to incorrect assumptions for zero transmission. Similarly, the application of a KBr correction for silicates is also problematic. Recent measurements by Colangeli indicate minimal matrix effects for various silicates. The application of a KBr correction of silicates has never been universally accepted in the way it has for SiC.

The 11.3 μm narrow feature often attributed to crystalline olivine would appear at 11.0 μm if the corrected spectrum were to be believed. Finally, for virtually all spectra obtained from dispersions, the derived mass extinction coefficients may be incorrect, due to the majority of the particles in the dispersions being "opaque". Thin film data on the other hand do not suffer from this problem.The new thin-film spectra and the uncorrected KBr-dispersion spectra to re-identify to 11.4 μm feature seen in the spectra of carbon-rich AGB stars.

Chapter 6

Magnetic and Dielectric Properties

MAGNETIC SUSCEPTIBILITY

The intensity of magnetization, I, is related to the strength of the inducing magnetic field, H, through a constant of proportionality,k, known as the magnetic susceptibility.

$$I = kH$$

The magnetic susceptibility is a unitless constant that is determined by the physical properties of the magnetic material. It can take on either positive or negative values. Positive values imply that the induced magnetic field,I, is in the same direction as the inducing field, H. Negative values imply that the induced magnetic field is in the opposite direction as the inducing field. In magnetic prospecting, the susceptibility is the fundamental material property whose spatial distribution we are attempting to determine. In this sense, magnetic susceptibility is analogous to density in gravity surveying.

INDUCED MAGNETIZATION

The nature of material magnetization is in general complex, governed by atomic properties, and well beyond the scope of this series of notes. Suffice it to say, there are three types of magnetic materials: paramagnetic, diamagnetic, and ferromagnetic.

Diamagnetism

Discovered by Michael Faraday in 1846. This form of magnetism is a fundamental property of all materials and is caused by the alignment of magnetic moments associated with orbital electrons* in the presence of an external magnetic field. For those elements with no unpaired electrons in their outer electron shells, this is the only form of magnetism observed. The susceptibilities of diamagnetic materials are relatively small and negative. Quartz and salt are two common diamagnetic earth materials.

Paramagnetism

This is a form of magnetism associated with elements that have an odd number of electrons in their outer electron shells. Paramagnetism is associated with the alignment of electron spin directions in the presence of an external magnetic field. It can only be observed at relatively low temperatures. The temperature above which paramagnetism is no longer observed is called the Curie Temperature. The susceptibilities of paramagnetic substances are small and positive.

Ferromagnetism

This is a special case of paramagnetism in which there is an almost perfect alignment of electron spin directions within large portions of the material referred to as domains. Like paramagnetism, ferromagnetism is observed only at temperatures below the Curie temperature. There are three varieties of ferromagnetism.

Pure Ferromagnetism

The directions of electron spin alignment within each domain are almost all parallel to the direction of the external inducing field. Pure ferromagnetic substances have large (approaching 1) positive susceptibilities. Ferromagnetic minerals do not exist, but iron, cobalt, and nickel are examples of common ferromagnetic elements.

Antiferromagnetism

The directions of electron alignment within adjacent domains are opposite and the relative abundance of domains with each spin direction is approximately equal. The observed magnetic intensity for the material is almost zero. Thus, the susceptibilities of antiferromagnetic materials are almost zero. Hematite is an antiferromagnetic material.

Ferromagnetism

Like antiferromagnetic materials, adjacent domains produce magnetic intensities in opposite directions. The intensities associated with domains polarized in a direction opposite that of the external field, however, are weaker. The observed magnetic intensity for the entire material is in the direction of the inducing field but is much weaker than that observed for pure ferromagnetic materials. Thus, the susceptibilities for ferromagnetic materials are small and positive. The most important magnetic minerals are ferromagnetic and include magnetite, titanomagnetite, ilmenite, and pyrrhotite.

VOLUME SUSCEPTIBILITY

The volume magnetic susceptibility, represented by the symbol (often simply , sometimes — magnetic, to distinguish from the electric susceptibility), is defined by the relationship

$$M = \chi_v H$$

where, in SI units,

M is the magnetization of the material (the magnetic dipole moment per unit volume), measured in amperes per meter, and

H is the magnetic field strength, also measured in amperes per meter. The magnetic induction B is related to H by the relationship,

$$B = \mu_0(H + M) = \mu_0(1 + \chi_v)H = \mu H$$

where μ_0 is the magnetic constant, and $(1 + \chi_v)$ is the relative permeability of the material. The magnetic susceptibility ÷v and the magnetic permeability μ are related by the following formula:

$$\mu = \mu_0(1 + \chi_v)$$

Sometimes an auxiliary quantity, called intensity of magnetization and measured in tesla, is defined as

$$I = \mu_0 M.$$

This allows an alternative description of all magnetization phenomena in terms of the quantities I and B, as opposed to the commonly used M and H.

CONVERSION BETWEEN SI AND CGS UNITS

Note that these definitions are according to SI conventions. However, many tables of magnetic susceptibility give cgs values (often denoted by "emu" or "e.m.u.", short for electromagnetic unit) that rely on a different definition of the permeability of free space:

$$B^{cgs} = H^{cgs} + 4\pi M^{cgs} = (1 + 4\pi\chi_v^{cgs})H^{cgs}$$

The dimensionless cgs value of volume susceptibility is multiplied by 4ð to give the dimensionless SI volume susceptibility value:

$$\chi_v^{SI} = 4\pi\chi_v^{cgs}$$

For example, the cgs volume magnetic susceptibility of water at 20°C is -7.19×10^{-7} which is -9.04×10^{-6} using the SI convention.

MASS AND MOLAR SUSCEPTIBILITY

There are two other measures of susceptibility, the mass magnetic susceptibility (÷mass or ÷g, sometimes ÷m), measured in $m^3 \cdot kg^{-1}$ in SI or in $cm^3 \cdot g^{-1}$ in cgs and the molar magnetic susceptibility (÷mol) measured in $m^3 \cdot mol^{-1}$ (SI) or $cm^3 \cdot mol^{-1}$ (cgs) that are defined below, where ñ is the density in $kg \cdot m^{-3}$ (SI) or $g \cdot cm^{-3}$ (cgs) and M is molar mass in $kg \cdot mol^{-1}$ (SI) or $g \cdot mol^{-1}$ (cgs).

÷mass = ÷v/η

÷mol = M÷mass = M÷v/η

SIGN OF SUSCEPTIBILITY

If ÷ is positive, then (1+÷v) > 1 (or, in cgs units, (1+4ð÷v) > 1) and the material can be paramagnetic, ferromagnetic, ferrimagnetic, or antiferromagnetic. In this case, the magnetic field is strengthened by the presence of the material. Alternatively, if ÷ is negative, then (1+÷v) < 1 (or, in cgs units, (1+4ð÷v) < 1), and the material is diamagnetic. As a result, the magnetic field is weakened in the presence of the material.

DETERMINATION OF SUSCEPTIBILITY

Volume magnetic susceptibility is measured by the force change felt upon the application of a magnetic field gradient. Early measurements were made using the Gouy balance where a sample is hung between the poles of an electromagnet. The change in weight when the electromagnet is turned on is proportional to the susceptibility.

Today, high-end measurement systems use a superconductive magnet. An alternative is to measure the force change on a strong compact magnet upon insertion of the sample. This system, widely used today, is called the Evan's balance. For liquid samples, the susceptibility can be measured from the dependence of the NMR frequency of the sample on its shape or orientation.

Tensor Susceptibility

The magnetic susceptibility of most crystals is not a scalar. Magnetic response M is dependent upon the orientation of the sample and can occur in directions other than that of the applied field H. In these cases, volume susceptibility is defined as a tensor,

$$Mi = \div ijHj$$

where i and j refer to the directions (e.g., x, y and z in Cartesian coordinates) of the applied field and magnetization, respectively. The tensor is thus rank 2, dimension (3,3) describing the component of magnetization in the i-th direction from the external field applied in the j-th direction.

Differential Susceptibility

In ferromagnetic crystals, the relationship between M and H is not linear. To accommodate this, a more general definition of differential susceptibility is used,

$$c^{d}_{ij} = \frac{\partial M_i}{\partial H_j}$$

where χ^{d}_{ij} is a tensor derived from partial derivatives of components of M with respect to components of H. When the coercivity of the material parallel to an applied field is the smaller of the two, the differential susceptibility is a function of the applied field and self interactions, such as the magnetic anisotropy. When the material is not saturated, the effect will be nonlinear and dependent upon the domain wall configuration of the material.

SUSCEPTIBILITY IN FREQUENCY DOMAIN

When the magnetic susceptibility is measured in response to an AC magnetic field (i.e. a magnetic field that varies sinusoidally), this is called AC susceptibility. AC susceptibility (and the closely-related "AC permeability") are complex quantities, and various phenomena (such as resonances) can be seen in AC susceptibility that cannot in constant-field (DC) susceptibility. In particular, when an ac-field is applied perpendicular to the detection direction (called the "transverse susceptibility" regardless of the frequency), the effect has a peak at the ferromagnetic resonance frequency of the material with a given static applied field. Currently, this effect is called the microwave permeability or network ferromagnetic resonance in the literature. These results are sensitive to the domain wall configuration of the material and eddy currents.

MAGNETOSTATIC FIELDS IN MATTER

MAGNETIZATION

Any macroscopic object consists of many atoms or molecules, each having electric charges in motion. With each electron in an atom or molecule we can associate a tiny magnetic dipole moment (due to its spin).

Ordinarily, the individual dipoles cancel each other because of the random orientation of their direction. However, when a magnetic field is applied, a net alignment of these magnetic dipoles occurs, and the material becomes magnetized.

The state of magnetic polarization of a material is described by the parameter M which is called the magnetization of the material and is defined as:

M = magnetic dipole moment per unit volume

In some material the magnetization is parallel to *B* These materials are called paramagnetic. In other materials the magnetization is opposite to *B* These materials are called diamagnetic. A third group of materials, also called Ferro magnetic materials, retain a substantial magnetization indefinitely after the external field has been removed.

PARAMAGNETISM

Consider a rectangular current loop, with sides s1 and s2, located in a uniform magnetic field, pointing along the z axis. The magnetic dipole moment of the current loop makes an angle **è** with the z axis.

The magnetic forces on the left and right sides of the current loop have the same magnitude but point in opposite directions.

The net force acting on the left and right side of the current loop is therefore equal to zero. The force on the top and bottom part of the current loop also have the same magnitude and point in opposite directions. However since these forces are not collinear, the corresponding torque is not equal to zero. The torque generated by magnetic forces acting on the top and the bottom of the current loop is equal to

$$\bar{N} = \sum \bar{r} \times \bar{F} = \frac{s_1}{2} F \sin\theta \hat{i} + \frac{s_1}{2} F \sin\theta \hat{i}$$

$$= s_i F \sin\theta \hat{i}$$

The magnitude of the force F is equal to:

$$F = \left| \int_{Line} d\bar{l} \times \bar{B} \right| = I s_2 B$$

Therefore, the torque on the current loop is equal to:

$$\bar{N} = s_1 s_2 IB \sin\theta \hat{i} = mB \sin\theta \hat{i} = \bar{m} \times \bar{B}$$

where $\bar{m}$ is the magnetic dipole moment of the current loop. As a result of the torque on the current loop, it will rotate until its dipole moment is aligned with that of the external magnetic field.

In atoms we can associate a dipole moment with each electron (spin). An external magnetic field will line up the dipole moment of the individual electrons (where not excluded by the Pauli principle). The induced magnetization is therefore parallel to the direction of the external magnetic field. It is this mechanism that is responsible for paramagnetism.

In a uniform magnetic field the net force on any current loop is equal to zero:

$$\bar{F} = I\oint[d\bar{l} \times \bar{B}] = I\left[\oint d\bar{l}\right] \times \bar{B} = 0$$

since the line integral of $d\bar{l}$ is equal to zero around any closed loop.

If the magnetic field is non-uniform then, in general, there will be a net force on the current loop. Consider an infinitesimal small current square of side Å, located in the yz plane and with a current flowing in a counter-clockwise direction. The force acting on the current loop is the vector sum of the forces acting on each side:

$$\bar{F} \; I\left[\int_{\text{Side1}} d\bar{l} \times \bar{B} + \int_{\text{Side2}} d\bar{l} \times \bar{B} + \int_{\text{Side3}} d\bar{l} \times \bar{B} + \int_{\text{Side4}} d\bar{l} \times \bar{B}\right]$$

$$= I\left[\int_{\dot{u}}^{s}\left(\hat{j} \times \bar{B}(0, y, 0)\right)dy + \int_{\dot{u}}^{s}\left(\hat{k} \times \bar{B}(0, \varepsilon, z,)\right)dz,\right.$$

$$\left. + \int_{s}^{\dot{u}}\left(\hat{j} \times \bar{B}(0, y, \varepsilon)\right)dy + \int_{s}^{\dot{u}}\left(\hat{k} \times \bar{B}(0, 0, z)\right)dz\right]$$

$$= I\left[\int_{\dot{u}}^{s}\left(\hat{j} \times \left[\bar{B}(0, y, 0) - \bar{B}(0, y, \varepsilon)\right]\right)dy\right.$$

$$\left. + \int_{\dot{u}}^{s}\left(\hat{k} \times \left[\bar{B}(0, \varepsilon, z) - \bar{B}) - 0, 0, z)\right]\right)dz\right]$$

$$= I\left[-\int_{\dot{u}}^{s}\left(\hat{j} \times \varepsilon \left.\frac{\partial \bar{B}}{\partial z}\right|_{(\dot{u}, y, 0)}\right)dy + \int_{\dot{u}}^{s}\left(\hat{k} \times \varepsilon \left.\frac{\partial \bar{B}}{\partial y}\right|_{(\dot{u}, \dot{u}, z)}\right)dz\right]$$

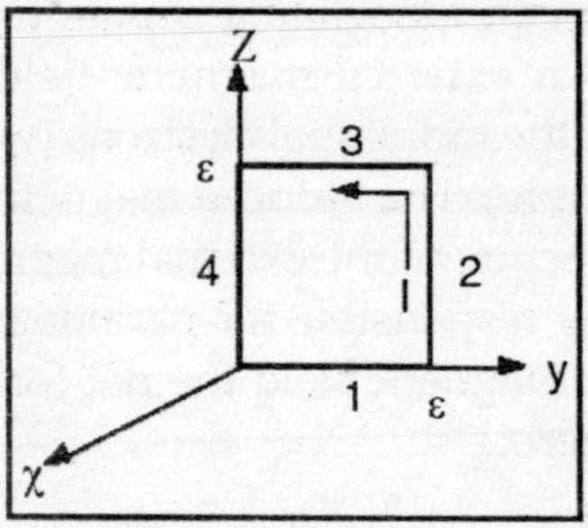

Fig. Infinitesimal Square Current Loop

In this derivation we have used a first-order Taylor expansion of $\bar{B}$:

$$= \bar{B}(0,0,z) + \varepsilon \frac{\partial \bar{B}}{\partial y}\bigg|_{(\dot{u},\dot{u},z)}$$

and,

$$\bar{B}(0,y,\varepsilon) = \bar{B}(0,y,0) + \varepsilon \frac{\partial \bar{B}}{\partial z}\bigg|_{(\dot{u},y,\ddot{u})}$$

Assuming that the current loop is so small that the derivatives of are constant over the boundaries of the loop we can evaluate the integrals and obtain for the total force:

$$\bar{F} = I\varepsilon^2 \left[\hat{k} \times \frac{\partial \bar{B}}{\partial y} - \hat{j} \times \frac{\partial \bar{B}}{\partial z} \right]$$

$$= m\left[\left(\frac{\partial B_{\perp}}{\partial y} \hat{j} - \frac{\partial B_y}{\partial y} \hat{i} \right) - \left(\frac{\partial b_z}{\partial z} \hat{i} - \frac{\partial B_{\perp}}{\partial y} \hat{k} \right) \right]$$

$$= m\left[\frac{\partial B_{\perp}}{\partial x} \hat{i} + \frac{\partial B_{\perp}}{\partial y} \hat{j} + \frac{\partial B_{\perp}}{\partial z} \hat{k} \right]$$

where m is the magnetic dipole moment of the current loop. In this derivation we have used the fact that the divergence of $\bar{B}$ is equal to zero for any magnetic field and this requires that,

$$\frac{\partial B_{\perp}}{\partial x} = \frac{\partial B_y}{\partial y} - \frac{\partial B_z}{\partial z}$$

The magnetic dipole moment m of the current loop is equal to,

$$\overline{m} = m\hat{i}$$

Therefore, the equation for the force acting on the current loop can be rewritten in terms of $\overline{m}$:

$$\overline{F} = \frac{\partial}{\partial x}(mB_{\perp})\hat{i} + \frac{\partial}{\partial y}(mB_{\perp})\hat{j} + \frac{\partial}{\partial z}(mB_{\perp})\hat{k}$$

$$= \overline{\nabla}(\overline{m} \bullet \overline{B})$$

Any current loop can be build up of infinitesimal current loops and therefore,

$$\overline{F} = \overline{\nabla}(\overline{m} \bullet \overline{B})$$

for any current loop.

DIAMAGNETISM

Consider a very classical picture of a Hydrogen atom consisting of an electron revolving in a circular orbit of radius r around a nucleus.

Suppose that the velocity of the electron is equal to v. Since the velocity of the electron is very high, the revolving electron looks like a steady current of magnitude,

$$I = \frac{e}{T} = \frac{ev}{2\pi r}$$

The direction of the current is in a direction opposite to that of the electron.

The dipole moment of this current is equal to,

$$\overline{m} = -\frac{1}{2}evr\,\hat{k}$$

If the atom is placed in a magnetic field, it will be subject to a torque. However, it is very difficult to tilt the entire orbit. Instead the electron will try to reduce its torque by changing its velocity. With no magnetic field present, the velocity of the electron can be obtained by requiring that the centripetal force is sustained by just the electric force:

$$\frac{1}{4\pi\varepsilon_{ü}}\frac{e^2}{r^2} = m_e\frac{v^2}{r}$$

In a magnetic field, the centripetal force will be sustained by both the electric and the magnetic field:

$$\frac{1}{4\pi\varepsilon_{ü}}\frac{e^2}{r^2}+ev'B=m_e\frac{v'^2}{r}$$

Here we have assumed that the magnetic field is pointing along the positive z axis (in a direction opposite to the direction of the magnetic dipole moment). We have also assumed that the size of the orbit (r) does not change when the magnetic field is applied.

Combining the last two equations we obtain,

$$m_e\frac{v^2}{r}+ev'B=m_e\frac{v'^2}{r}$$

or,

$$ev'B=\frac{m_e}{r}\left(v'^2-v^2\right)=\frac{m_e}{r}(v'+v)(v'-v)$$

Assuming that the change in the velocity is small we can use the following approximations:

$$v'\cong v$$

and

$$v'-v\cong\Delta v$$

Therefore,

$$evB=2\frac{m_e}{r}v\Delta v$$

This equation shows that the presence of the magnetic field will increase the speed of the electron. An increase in the velocity of the electron will increase the magnitude of the dipole moment of the revolving electron. The change in is opposite to the direction of $\overline{B}$. If the electron would have been orbiting the other way, it would have been slowed down by the magnetic field. Again the change in the dipole moment is opposite to the direction of $\overline{B}$.

In the presence of an external magnetic field the dipole moment of each orbit will be slightly modified, and all these changes are anti-parallel to the external magnetic field. This is the magnetism that is responsible for diamagnetism. Diamagnetism is present in all materials, but is in general much weaker than paramagnetism. It can therefore only be observed in those materials where paramagnetism is not present.

MAGNETIZED FIELD OBJECT

Consider a magnetized material with magnetization $\bar{M}$. The associated vector potential $\bar{A}$ isp equal to,

$$\bar{A} = \frac{\mu_{\ddot{u}}}{4\pi} \int \frac{\bar{M} \times \Delta\hat{r}}{\Delta r^2} d\tau$$

Following the same procedure used to calculate the electrostatic potential of a polarized material, we obtain for $\bar{A}$:

$$\bar{A} = \frac{\mu_{\ddot{u}}}{4\pi} \int_{\text{Surface}} \frac{\bar{\nabla} \times \bar{M}}{\Delta r} d\tau + \frac{\mu_{\ddot{u}}}{4\pi} \oint_{\text{Surface}} \frac{\bar{M} \times d\bar{a}}{\Delta r}$$

$$= \frac{\mu_{\ddot{u}}}{4\pi} \int_{\text{Surface}} \frac{\bar{J}\delta}{\Delta r} d\tau + \frac{\mu_{\ddot{u}}}{4\pi} \oint_{\text{Surface}} \frac{\bar{K}_{\delta}}{\Delta r}$$

where $\bar{J}_{\delta} = \bar{\nabla} \times \bar{M}$ is the bound volume current and $\bar{K}_{\delta} = \bar{M} \times \hat{n}$ is the bound surface current. If the material has a uniform magnetization then the bound volume current is zero. The field produced by a magnetized object is equal to the field produced by the bound currents. Consider a uniformly magnetized thin slab of material of thickness t. The material can be cut up into tiny current loops. If each current loop has an area a then the dipole moment due to a surface current I is equal to

$$\bar{m}_i = Ia\hat{k}$$

The volume of the current loop is at and therefore its dipole moment must be equal to

$$\bar{m}_i = Mat\hat{k}$$

This requires that the surface current of the current loop is equal to

$$I = Mt$$

Since the magnetization is uniform, the current in each of the current loops will be constant and flowing in the same direction. Therefore, all volume currents cancel, and the only current remaining will be a surface current, flowing on the surface of the material. The current flowing on the surface of the material will be equal to the current in each of the current loops.

Therefore, the current density on the surface is equal to,

$$K = \frac{I}{t} = M$$

In vector notation:

$$\bar{K} = \bar{M} \times \hat{n}$$

This equation is also consistent with the fact that there is no current flowing on the top and bottom surfaces (where).

$$\bar{M} \times \hat{n} = 0$$

AUXILIARY FIELD H

To calculate the magnetic field produced by a system containing magnetized materials we have to use the following form of Ampere's law:

$$\frac{1}{\mu_{\ddot{u}}}(\bar{\nabla} \times \bar{B}) =$$

This equation can be rewritten as,

$$\bar{\nabla} \times \left(\frac{1}{\mu_{\ddot{u}}} \bar{B} - \bar{M} \right) =$$

The quantity in parenthesis is called the H-field,

$$\bar{H} = \frac{1}{\mu_{\ddot{u}}} \bar{B} - \bar{M}$$

$\bar{H}$ plays a role in magnetostatics analogous to $\bar{D}$ in electrostatics. Ampere's law in terms of $\bar{H}$ reads

$$\bar{\nabla} \times \bar{H} =$$

and,

$$\oint \bar{H} \bullet dl$$

Ampere's law for $\bar{H}$ tells us that the curl of $\bar{H}$ is equal to the free current density. However, a knowledge of the free current density is not sufficient to determine $\bar{H}$.

The Helmholtz theorem shows that besides knowing the curl of a vector function, we also need to know the divergence of that vector function before it is uniquely defined. Although the divergence of $\bar{B}$ is zero for any magnetic field (and therefore Ampere's law for $\bar{B}$ defines $\bar{B}$ uniquely) the divergence of $\bar{H}$ is not necessarily zero:

$$\bar{\nabla} \bullet \bar{H} = \bar{\nabla} \bullet \left(\frac{1}{\mu_{\ddot{u}}} \bar{B} - \bar{M} \right)$$

$$= \frac{1}{\mu_{\ddot{u}}} \bar{\nabla} \bullet \bar{B} - \bar{\nabla} \bullet \bar{M} = -\bar{\nabla} \bullet \bar{M}$$

Therefore, only for those systems where $\bar{\nabla} \bullet \bar{M} = 0$ can we use Ampere's law for $\bar{H}$ directly to calculate $\bar{H}$. The divergence of $\bar{H}$ will be zero only for systems with cylindrical, plane, solenoidal, or toroidal symmetry.

The $\bar{H}$ field is a quantity that is used in the laboratory more often that the $\bar{B}$ field. This is a result of the dependence of $\bar{H}$ on only the free currents (which are easy to control). The $\bar{B}$ field depends both on the free and on the bound currents, and thus requires a detailed knowledge of the magnetic properties of the materials used.

In electrostatics, the electric field can be obtained immediately from the potential difference (which is easy to control).

The electric displacement $\bar{D}$ depends only on the free charge distribution, but in most cases a direct measurement of the free charge distribution is very difficult to carry out. Therefore, in electrostatics the electric field is in most cases a more useful parameter then the electric displacement $\bar{D}$.

Linear Media

In paramagnetic and diamagnetic materials, the magnetization is maintained by the external magnetic field. The magnetization disappears when the field is removed. Most paramagnetic and diamagnetic materials are linear; that is their magnetization is proportional to the $\bar{H}$ field:

$$\bar{M} = \chi_{vc}\bar{H}$$

The constant of proportionality is χ_{vc} called the magnetic susceptibility of the material. In vacuum the magnetic susceptibility is zero. In a linear medium, there is linear relation between the magnetic field and the $\bar{H}$ field:

$$\bar{B} = \mu_0(\bar{H} + \bar{M}) = \mu_0(1 + \chi_{vc})\bar{H} = \mu\bar{H}$$

where μ is called the permeability of the material. The permeability of free space is equal to $\mu_{\ddot{u}}$.

The linear relation between $\bar{H}$ and $\bar{B}$ does not automatically imply that the divergence of $\bar{H}$ is zero. The divergence of $\bar{H}$ will only be equal to zero inside a linear material, but will be non-zero at the interface between two materials of different permeability. Consider for example the interface between a linear material and vacuum. The surface integral of $\bar{M}$ across the surface of the Gaussian pillbox shown in Figure 6.7 is definitely not equal to zero. According to the divergence theorem the surface integral of $\bar{M}$ is equal to the volume integral of $\bar{\nabla} \bullet \bar{M}$:

$$\oint_{\text{Surface}} \bar{M} \bullet d\bar{a} = \int_{\text{Volume}} (\bar{\nabla} \bullet \bar{M})\, d\tau$$

Fig. Interface of Linear Materials

Therefore, if the surface integral of $\bar{M}$ is not equa! to zero, the divergence of $\bar{M}$ can not be zero everywhere.

Nonlinear Media

The best known nonlinear media are the ferromagnetic materials. Feiromagnetic materials do not require external fields to sustain their magnetization (therefore, the magnetization definitely depends in a nonlinear way on the field). The magnetization in ferromagnetic materials involves the alignment of the dipole moments associated with the spin of unpaired electrons. The difference between ferromagnetic materials and paramagnetic materials is that in ferromagnetic materials the interaction between nearby dipoles makes them want to point in the same direction, even when the magnetic field is removed. However, the alignment occurs in relative small patches, called domains. When a ferromagnetic material is not located in a magnetic field, the dipole moments of the various domains are not aligned, and the material as a whole is not magnetized. When the ferromagnetic material is put into a magnetic field, the boundaries of the domain parallel to the field will increase at the expense of neighboring boundaries. If the field is strong enough, one domain takes over the entirely, and the ferromagnetic material is said to be saturated (all unpaired electrons are aligned and therefore the magnetization reaches a maximum value).

The magnetic susceptibility of ferromagnetic materials is around 103 (roughly eighth orders of magnitude larger than the susceptibility of paramagnetic materials). When the magnetic field is removed some magnetization remains (and we have created a permanent magnet). For any ferromagnetic material, the magnetization depends not only on the applied magnetic field but also on the magnetization history. The alignment of dipoles in a ferromagnet can be destroyed by random thermal motion. The destruction of the alignment occurs at a precise temperature (called the Curie point). When a ferromagnetic material is heated above its Curie temperature it becomes paramagnetic.

Principles of Magnetic Susceptibility

Magnetic susceptibility measurements are a non-destructive and cost effective method of determining the presence of iron-bearing minerals within the sediments. The whole core, or individual sediment samples, are exposed to an external magnetic field which causes the sediments to become magnetized according to the amount of Fe-bearing minerals present in the samples. Magnetic susceptibility is a measure of the ease with which particular sediments are magnetized when subjected to a magnetic field. The ease of magnetization is ultimately related to the concentration and composition (size, shape and mineralogy) of magnetizable material contained within the sample. Any sediment core possessing downcore variation per unit volume in the concentration and composition of magnetizable minerals will yield a MS curve reflecting these changes.

In our system, whole cores are moved incrementally (generally in 1cm) by a track motor through a susceptibility loop (of varying size) in which a magnetic field is generated and which magnetizes the sample susceptible substances (minerals or mineraloids) within the sediment. Samples that are rich, per unit volume, in magnetizable substances will yield high readings. Samples that are poor in magnetizable substances, or contain diamagnetic minerals, will yield lower or negative values Magnetizable minerals include the ferromagnetic minerals (strongly magnetizable) and any of the paramagnetic (moderately magnetizable) minerals and other substances.

The former include magnetite, hematite, iron titanium oxides, pyrrhotite, maghemite, greigite and goethite, minerals capable of acquiring remnant magnetization and useful for paleomagnetic studies. The latter include a broad array of substances all of which contain Fe^{2+}, Fe^{3+}, or Mn^{2+} ions. These paramagnetic minerals may include clay minerals (chlorite, smectite and glauconite), iron and manganese carbonates (siderite, rhodochrosite), ferromagnesian silicates (olivine, amphiboles, pyroxenes, etc.), as well as a variety of ferric-oxyhydroxide mineraloids.

LRC Magnetics Instrumentation

LRC maintains a Bartington Instruments MS2 magnetic susceptibility meter, sensors MS2B and C, as part of the multi-sensor track logging system (MSTL), as well as MS2 field survey equipment.. The system is a non-destructive method of obtaining high resolution (1cm) measurement of magnetic susceptibility (MS) from whole cores (other systems are designed to work with split cores). The system also can be used to measure single sample susceptibility. The MSTL forms the first step in the initial core studies, prior to splitting the cores. For whole core measurements, we have designed a maple magnetic susceptibility track. This track is 3.2 meters long and includes a Kevlar timing belt and a PVC boat which can accommodate a core up to 7.5 cm in diameter.

The MS2C core logging sensor is attached to the track. This sensor is 100 mm in diameter and provides high precision volume susceptibility measurements on whole cores. It is suitable for measuring any type of peat, lake or marine sediment core, provided the core is not metal clad.

The instrument has a precision of 1 × 10-6 cgs and can log cores at intervals down to ca. 20 mm. The single sample sensor MS2B is designed for use with 1" and 7/8" cubic boxes. Sample insertion and removal is facilitated by a hand operated moving platen.

This sensor is used for measurements on standard paleomagnetic samples in studies of rock and mineral magnetism and anisotropy of susceptibility. The dual frequency facility allows confident identification of ferrimagnetic stable single domain transition. It is accurate to 2×10^{7} cgs for both high and low frequencies.

FERRIMAGNETISM

A specific type of ordering in a system of magnetic moments or the magnetic behaviour resulting from such order. In some magnetic materials the magnetic ions in a crystal unit cell may differ in their magnetic properties. This is clearly so when some of the ions are of different species. It is also true for similar ions occupying crystallographically inequivalent sites. Such ions differ in their interactions with other ions, because the dominant exchange interaction is mediated by the neighboring nonmagnetic ions.

They also experience different crystal electric fields, and these affect the magnetic anisotropy of the ion. A collection of all the magnetic sites in a crystal with identical behaviour is referred to as a magnetic sublattice. A material is said to exhibit ferrimagnetic order when, first, all moments on a given sublattice point in a single direction and, second, the resultant moments of the sublattices lie parallel or antiparallel to one another. The notion of such an order is due to L. Néel, who showed in 1948 that its existence would explain many of the properties of the magnetic ferrites.

In general, there is a net moment, the algebraic sum of the sublattice moments, just as for a normal ferromagnet. However, its variation with temperature rarely exhibits the very simple behaviour of the normal ferromagnet. For example, in some materials, as the temperature is raised over a certain range, the magnetization may first decrease to zero and then increase again. Ferrimagnets can be expected, in their bulk properties, measured statically or at low frequencies, to resemble ferromagnets with unusual temperature characteristics.

Ferrimagnetic materials are like ferromagnets in that they hold a spontaneous magnetization below the Curie temperature, and show no magnetic order (are paramagnetic) above this temperature. However, there is sometimes a temperature below the Curie temperature at which the two sublattices have equal moments, resulting in a net magnetic moment of zero; this is called the magnetization compensation point. This compensation point is observed easily in garnets and rare earth - transition metal alloys (RE-TM). Furthermore, ferrimagnets may also exhibit an angular momentum compensation point at which the angular momentum of the magnetic sublattices is compensated. This compensation point is a crucial point for achieving high speed magnetization reversal in magnetic memory devices .

Ferrimagnetism is exhibited by ferrites and magnetic garnets. The oldest-known magnetic substance, magnetite (iron(II,III) oxide; Fe_3O_4), is a ferrimagnet; it was originally classified as a ferromagnet before Néel's discovery of ferrimagnetism and antiferromagnetism .

Some ferrimagnetic materials are YIG (yttrium iron garnet) and ferrites composed of iron oxides and other elements such as aluminum, cobalt, nickel, manganese and zinc.

PROPERTIES OF FERRIMAGNETISM

Ferrimagnetic materials have high resistivity and have

anisotropic properties. The anisotropy is actually induced by an external applied field. When this applied field aligns with the magnetic dipoles it causes a net magnetic dipole moment and causes the magnetic dipoles to precess at a frequency controlled by the applied field, called Larmor or precession frequency.

As a particular example, a microwave signal circularly polarized in the same direction as this precession strongly interacts with the magnetic dipole moments; when it is polarized in the opposite direction the interaction is very low. When the interaction is strong the microwave signal can pass through the material.

This directional property is used in the construction of microwave devices like isolators, circulators and gyrators. Ferrimagnetic materials are also used to produce optical isolators and circulators.

FERRITE

Any of the class of magnetic oxides. Typically the ferrites have a crystal structure which has more than one type of site for the cations. Usually the magnetic moments of the metal ions on sites of one type are parallel to each other, and antiparallel to the moments on at least one site of another type. Thus ferrites exhibit ferrimagnetism.

There are three important classes of commercial ferrites. One class has the spinel structure, with the general formula $M^{2+}Fe_2^{3+}O_4$, where M^{2+} is a divalent metal ion. So-called linear ferrites used in inductors and transformers are made of Mn and Zn (for frequencies up to 1 MHz) and Ni and Zn (for frequencies greater than 1 MHz). MgMn ferrites are used in microwave devices such as isolators and circulators. Until the late 1970s, ferrites with square loop shapes held a dominant position as computer memory-core elements, but these gave way to semiconductors.

The second class of commercially important ferrites have the garnet structure, with the formula M33+Fe53+O12, where M3+ is a rare-earth or yttrium ion. Yttrium-based garnets are used in microwave devices. Thin monocrystalline films of complex garnets have been developed for bubble domain memory devices.

The third class of ferrites has a hexagonal structure, of the $M^{2+}Fe_{12}{}^{3+}O_{19}$ magnetoplumbite type, where M^{2+} is usually Ba, Sr, or Pb. Because of their large magnetocrystalline anisotropy, the hexagonal ferrites develop high coercivity and are an important member of the permanent magnet family. Another magnetic oxide, Fe_2O_3, also has the spinel structure, but has no divalent cations. It is the most commonly used material in the preparation of magnetic recording tapes.

PROPERTIES OF FERRITE

Ferrites are usually non-conductive ferrimagnetic ceramic compounds derived from iron oxides such as hematite (Fe_2O_3) or magnetite (Fe_3O_4) as well as oxides of other metals. Ferrites are, like most other ceramics, hard and brittle. In terms of the magnetic properties, ferrites are often classified as "soft" and "hard" which refers to their low or high coercivity of their magnetism, respectively.

Soft ferrites

Ferrites that are used in transformer or electromagnetic cores contain nickel, zinc, or manganese compounds. They have a low coercivity and are called soft ferrites. Because of their comparatively low losses at high frequencies, they are extensively used in the cores of Switched-Mode Power Supply (SMPS) and RF transformers and inductors. A common ferrite, abbreviated "MnZn," is composed of the oxides of manganese and zinc.

Hard ferrites

In contrast, permanent ferrite magnets (or "hard ferrites"), which have a high remanence after magnetization, are composed of iron and barium or strontium oxides. In a magnetically saturated state they conduct magnetic flux well and have a high magnetic permeability. This enables these so-called ceramic magnets to store stronger magnetic fields than iron itself. They are the most commonly used magnets in radios. The maximum magnetic field B is about 0.35 tesla and the magnetic field strength H is about 30 to 160 kiloampere turns per meter (400 to 2000 oersteds).

PRODUCTION OF FERRITE MATERIALS

Ferrites are produced by heating an intimate mixture of powdered precursors are heated and pressed in a mold. During the heating process, calcination of carbonates occurs:

$$MCO_3 \rightarrow MO + CO_2$$

The oxides of barium and strontium are typically supplied as their carbonates, $BaCO_3$ or $SrCO_3$. The resulting mixture of oxides undergoes sintering. Afterwards the cooled product is milled to particles smaller than 2 μm in order to produce Weiss domains in the size of one particle. Next the powder is pressed into a shape, dried, and re-sintered. The shaping may be performed in an external magnetic field, in order to achieve a preferred orientation of the particles (anisotropy).Small and geometrically easy shapes may be produced with dry pressing.

However, in such a process small particles may agglomerate and lead to poorer magnetic properties compared to the wet pressing process. Direct calcination and sintering without re-milling is possible as well but leads to poor magnetic properties.Electromagnets are pre-sintered as well (pre-reaction), milled and pressed. However, the sintering takes place in a specific atmosphere, for instance one with an oxygen shortage). The chemical composition and especially the structure vary strongly between the precursor and the sintered product.

USES OF FERRITE

Ferrite cores are used in electronic inductors, transformers, and electromagnets where the high electrical resistance of the ferrite leads to very low eddy current losses. They are commonly seen as a lump in a computer cable, called a ferrite bead, which helps to prevent high frequency electrical noise (radio frequency interference) from exiting or entering the equipment.Early computer memories stored data in the residual magnetic fields of hard ferrite cores, which were assembled into arrays of core memory. Ferrite powders are used in the coatings of magnetic recording tapes.

One such type of material is iron (III) oxide. Ferrite particles are also used as a component of radar-absorbing materials or coatings used in stealth aircraft and in the expensive absorption tiles lining the rooms used for electromagnetic compatibility measurements.Most common radio magnets, including those used in loudspeakers, are ferrite magnets. Ferrite magnets have largely displaced Alnico magnets in these applications.It is a common magnetic material for electromagnetic instrument pickups, because of price and relatively high output. However, such pickups lack certain sonic qualities found in other pickups, such as those that use Alnico alloys or more sophisticated magnets.

Ferrite Devices

Electrical devices whose principle of operation is based upon the use and properties of ferrites, which are magnetic oxides. Ferrite devices are divided into two categories, depending on whether the ferrite is magnetically soft (low coercivity) or hard (high coercivity). Soft ferrites are used primarily as transformers, inductors, and recording heads, and in microwave devices. Since the electrical resistivity of soft ferrites is typically 106–1011 times that of metals, ferrite components have much lower eddy current losses and hence are used at frequencies generally above about 10 kHz. Hard ferrites are used in permanent-magnet motors, loudspeakers, and holding devices, and as storage media in magnetic recording devices. One type of soft ferrite, referred to as square-loop ferrites, was once deployed in huge quantities as cores for digital computer memories but has now been replaced by semiconductor integrated circuits.

ELECTRIC POLARIZATION

The response of individual electric charges to an electric field can be described with Newton's laws. The behaviour of a charged fluid has a field description that will not be discussed here. The behaviour of many solids in response to an electric field can be described remarkably well with the same kind of approximation that we made for stress and strain. If we think of stress as the perturbation applied to a sample and the strain as the response of the sample to the perturbation, we can say that in our model the response is proportional to the perturbation.

The response of many common solids to an applied electric field is such that the electric field inside the sample is less than the applied electric. It is as if the solid has produced its own charge distribution - become polarized so that it produces an electric field which adds, in opposition, to the applied field. Materials that do this are called dielectrics.

We take the polarization of the material to be the response in this case. It turns out that in many useful situations the polarization is proportional to the applied electric field. We need to define the polarization in a way that allows us to measure it. We follow tradition and define it as,

$$\vec{P} = \alpha\vec{E}$$

where the constant of proportionality, α, is called the dielectric polarizability.

This definition does not yet tell us how to measure the polarization and determine the size of α. That is accomplished by relating $\vec{P}$ to the internal electric field which reduces the applied electric field. Calling the internal electric field E_{BACK} we write,

$$E_{BACK} = \frac{P}{\varepsilon_0} = \frac{\alpha}{\varepsilon_0} E_{APPLIED}$$

where ε_0 is the usual electric constant,

$$\varepsilon_0 = 8.85 \times 10^{-12} \frac{Coul}{Newton/m^2}$$

With this definition, the polarization has units of electric dipole moment per unit volume. This is consistent with microscopic models of what happens inside a material to produce the observed electric field.

Switching to subscript notation again, and allowing the possibility that an electric field in one direction might produce a polarization in another (as is sometimes observed experimentally), we can write,

$$P_i = \alpha_{ij} E_i$$

where we use the Einstein Summation Convention, that a repeated index (in this case the repeated j index) is summed. Thus typically the ith component of polarizability is calculated as the sum of three contributions; one from each component of the applied field.

α_{ij} is called the polarizability tensor. It can be represented as a 3 × 3 array of 9 numbers.

$$\begin{pmatrix} \alpha_{11} & \alpha_{12} & \alpha_{13} \\ \alpha_{21} & \alpha_{22} & \alpha_{23} \\ \alpha_{31} & \alpha_{32} & \alpha_{33} \end{pmatrix}$$

In an isotropic material, the elements are zero unless $i = j$. Then the array has three non-zero elements, in a diagonal line from the upper left to the lower right corner. Moreover, in an isotropic material, there are no preferred directions.

This means that,

$$\alpha_{11} = \alpha_{22} = \alpha_{33} \equiv \alpha$$

For anisotropic materials, the diagonal elements are generally not equal to each other, and the off-diagonal elements are generally not zero. Two things are true, however:

- The array is symmetric, meaning that $\alpha_{ij} = \alpha_{ji}$. It is also said that the array is symmetric under the interchange of i and j.
- There exists a coordinate system in which the array has no off-diagonal elements and is said to be "diagonal." (There is actually a minimum of 3 such right-handed coordinate systems.) Even in this coordinate system, for anisotropic materials, the three diagonal elements cannot all be equal.

The problem of finding the values of the three α_{ij}'s in the diagonal array is called the "eigenvalue" problem. The three α_{ij}'s are called "eigen values," or eigenvalues, meaning unique values. The coordinate axes in which the array is diagonal are called "eigenvectors." The process of finding them is called "diagonalization" of the array (usually called a matrix).

Chapter 7

Superconductivity

PHENOMENON

Superconductivity is a phenomenon which occurs in certain materials and is characterized by the absence of electrical resistivity. Until recently, this phenomenon had been restricted to metals and alloys with transition temperatures of less than 23K. In 1986, superconductivity was discovered in a ceramic material. This precipitated an onrush of ceramic based superconductors with transition temperatures as high as 120K. The ceramic-based materials are commonly known as high temperature (HTc) superconductors while the metallic and alloy materials are called low temperature (LTc) superconductors. Currently, only low temperature superconductors are of interest to the magnet designer and manufacturer.

Superconductors are divided into two types depending on their characteristic behaviour in the presence of a magnetic field. Type I superconductors are comprised of pure metals, whereas Type II superconductors are comprised primarily of alloys or intermetallic compounds.

The critical temperature at which the resistance vanishes in a superconductor is reduced when a magnetic field is applied. The maximum field that can be applied to a superconductor at a particular temperature and still maintain superconductivity is call the critical field, or Hc. This field varies enormously between Type I and Type II superconductors. The maximum critical field (Hc) in any Type I superconductor is about 2000 Gauss (0.2 Tesla), but in Type

II materials superconductivity can persist to several hundred thousand Gauss (Hc2). At fields greater than Hc in a Type I superconductor and greater than Hc2 in a Type II superconductor, the conductor reverts to the normal state and regains its normal state resistance.

A Type I superconductor excludes the applied magnetic field from the centre of the sample by establishing circulating currents on its surface that counteract the applied field. Type II superconductors, however, permit the field to penetrate through the sample in quantized amounts of flux. These quanta are comprised of circulating vortices of current and the flux contained in the vortices.

The total flux in a vortex is 2×10^{-7} Gauss-cm^2. Great numbers of these vortices, or fluxoids as they are frequently called, can exist in a superconductor. For example, at a field intensity of 80 kilogauss (8 Tesla) there are 4×10^{11} fluxoids/cm^2. These fluxoids and their interactions with defects in the superconductor give rise to the high current carrying capabilities of superconducting magnets.

ROLE OF FLUX

Properties of superconducting materials are altered locally by the presence of defects in the materials. A fluxoid encompassing or adjacent to such a defect in the material has its energy altered and its free motion through the superconductor is inhibited. This phenomenon, known as flux pinning, causes a field gradient in the superconductor and gives rise to a net current in the material. In the absence of defects in a Type II superconductor, no bulk current can be conducted without a transition into the normally conducting resistive state.

Since the pinning force is small, fluxoids can be broken loose from their pinning centers, resulting in a net creep of the flux through a conductor as a function of time. This results in an effective voltage in a Type II superconductor. If the current density is low and the magnetic field is not intense, flux creep is insignificant and the induced voltage and effective resistance of the conductor will be essentially zero. At very high fields and high current densities, fluxoids will migrate rapidly, giving rise to a phenomenon called flux flow.

PROPERTIES OF SUPERCONDUCTORS

Most of the physical properties of superconductors vary from material to material, such as the heat capacity and the critical temperature, critical field, and critical current density at which superconductivity is destroyed.On the other hand, there is a class of properties that are independent of the underlying material. For instance, all superconductors have exactly zero resistivity to low applied currents when there is no magnetic field present. The existence of these "universal" properties implies that superconductivity is a thermodynamic phase, and thus possess certain distinguishing properties which are largely independent of microscopic details.

ZERO ELECTRICAL RESISTANCE

The simplest method to measure the electrical resistance of a sample of some material is to place it in an electrical circuit in series with a current source I and measure the resulting voltage V across the sample. The resistance of the sample is given by Ohm's law as,

$$R = \frac{V}{I}.$$

If the voltage is zero, this means that the resistance is zero and that the sample is in the superconducting state.

Superconductors are also able to maintain a current with no applied voltage whatsoever, a property exploited in superconducting electromagnets such as those found in MRI machines. Experiments have demonstrated that currents in superconducting coils can persist for years without any measurable degradation. Experimental evidence points to a current lifetime of at least 100,000 years, and theoretical estimates for the lifetime of a persistent current exceed the estimated lifetime of the universe.

In a normal conductor, an electrical current may be visualized as a fluid of electrons moving across a heavy ionic lattice. The electrons are constantly colliding with the ions in the lattice, and during each collision some of the energy carried by the current is absorbed by the lattice and converted into heat, which is essentially the vibrational kinetic energy of the lattice ions. As a result, the energy carried by the current is constantly being dissipated. This is the phenomenon of electrical resistance.

The situation is different in a superconductor. In a conventional superconductor, the electronic fluid cannot be resolved into individual electrons. Instead, it consists of bound pairs of electrons known as Cooper pairs. This pairing is caused by an attractive force between electrons from the exchange of phonons. Due to quantum mechanics, the energy spectrum of this Cooper pair fluid possesses an energy gap, meaning there is a minimum amount of energy ÄE that must be supplied in order to excite the fluid. Therefore, if ÄE is larger than the thermal energy of the lattice, given by kT, where k is Boltzmann's constant and T is the temperature, the fluid will not be scattered by the lattice. The Cooper pair fluid is thus a superfluid, meaning it can flow without energy dissipation.

In a class of superconductors known as Type II superconductors, including all known high-temperature superconductors, an extremely small amount of resistivity appears at temperatures not too far below the nominal superconducting transition when an electrical current is applied in conjunction with a strong magnetic field, which may be caused by the electrical current. This is due to the motion of vortices in the electronic superfluid, which dissipates some of the energy carried by the current. If the current is sufficiently small, the vortices are stationary, and the resistivity vanishes. The resistance due to this effect is tiny compared with that of non-superconducting materials, but must be taken into account in sensitive experiments. However, as the temperature decreases far enough below the nominal superconducting transition, these vortices can become frozen into a disordered but stationary phase known as a "vortex glass". Below this vortex glass transition temperature, the resistance of the material becomes truly zero.

Superconducting Phase Transition

In superconducting materials, the characteristics of superconductivity appear when the temperature T is lowered below a critical temperature Tc. The value of this critical temperature varies from material to material. Conventional superconductors usually have critical temperatures ranging from around 20 K (Kelvin) to less than 1 K. Solid mercury, for example, has a critical temperature of 4.2 K. As of 2001, the highest critical temperature found for a conventional superconductor is 39 K for magnesium diboride (MgB2), although this material displays enough exotic properties that there is doubt about classifying it as a "conventional" superconductor.

Cuprate superconductors can have much higher critical temperatures: YBa2Cu3O7, one of the first cuprate superconductors to be discovered, has a critical temperature of 92 K, and mercury-based cuprates have been found with critical temperatures in excess of 130 K. The explanation for these high critical temperatures remains unknown. Electron pairing due to phonon exchanges explains superconductivity in conventional superconductors, but it does not explain superconductivity in the newer superconductors that have a very high critical temperature.

The onset of superconductivity is accompanied by abrupt changes in various physical properties, which is the hallmark of a phase transition.

For example, the electronic heat capacity is proportional to the temperature in the normal (non-superconducting) regime. At the superconducting transition, it suffers a discontinuous jump and thereafter ceases to be linear. At low temperatures, it varies instead as T for some constant .

Behaviour of heat capacity (cv, blue) and resistivity (ñ, green) at the superconducting phase transition

The order of the superconducting phase transition was long a matter of debate. Experiments indicate that the transition is second-order, meaning there is no latent heat. Calculations in the 1970s suggested that it may actually be weakly first-order due to the effect of long-range fluctuations in the electromagnetic field. Only recently it was shown theoretically with the help of a disorder field theory, in which the vortex lines of the superconductor play a major role, that the transition is of second order within the type II regime and of first order (i.e., latent heat) within the type I regime, and that the two regions are separated by a tricritical point.

MEISSNER EFFECT

When a superconductor is placed in a weak external magnetic field H, the field penetrates the superconductor only a small distance λ, called the London penetration depth, decaying exponentially to zero within the bulk of the material. This is called the Meissner effect, and is a defining characteristic of superconductivity. For most superconductors, the London penetration depth is on the order of 100 nm.The Meissner effect is sometimes confused with the kind of diamagnetism one would expect in a perfect electrical conductor: according to Lenz's law, when a changing magnetic field is applied to a conductor, it will induce an electrical current in the conductor that creates an opposing magnetic field.

In a perfect conductor, an arbitrarily large current can be induced, and the resulting magnetic field exactly cancels the applied field.The Meissner effect is distinct from this because a superconductor expels all magnetic fields, not just those that are changing. Suppose we have a material in its normal state, containing a constant internal magnetic field. When the material is cooled below the critical temperature, we would observe the abrupt expulsion of the internal magnetic field, which we would not expect based on Lenz's law.The Meissner effect was explained by the brothers Fritz and Heinz London, who showed that the electromagnetic free energy in a superconductor is minimized provided

$$\nabla^2 H = \lambda^{-2} H$$

where H is the magnetic field and λ is the London penetration depth.

This equation, which is known as the London equation, predicts that the magnetic field in a superconductor decays exponentially from whatever value it possesses at the surface. The Meissner effect breaks down when the applied magnetic field is too large. Superconductors can be divided into two classes according to how this breakdown occurs. In Type I superconductors, superconductivity is abruptly destroyed when the strength of the applied field rises above a critical value Hc. Depending on the geometry of the sample, one may obtain an intermediate state consisting of regions of normal material carrying a magnetic field mixed with regions of superconducting material containing no field.

In Type II superconductors, raising the applied field past a critical value Hc1 leads to a mixed state in which an increasing amount of magnetic flux penetrates the material, but there remains no resistance to the flow of electrical current as long as the current is not too large. At a second critical field strength Hc2, superconductivity is destroyed. The mixed state is actually caused by vortices in the electronic superfluid, sometimes called fluxons because the flux carried by these vortices is quantized. Most pure elemental superconductors, except niobium, technetium, vanadium and carbon nanotubes, are Type I, while almost all impure and compound superconductors are Type II.

LONDON MOMENT

Conversely, a spinning superconductor generates a magnetic field, precisely aligned with the spin axis. The effect, the London moment, was put to good use in Gravity Probe B. This experiment measured the magnetic fields of four superconducting gyroscopes to determine their spin axes. This was critical to the experiment since it is one of the few ways to accurately determine the spin axis of an otherwise featureless sphere.

THEORIES

Since the discovery of superconductivity, great efforts have been devoted to finding out how and why it works. During the 1950s, theoretical condensed matter physicists arrived at a solid understanding of "conventional" superconductivity, through a pair of remarkable and important theories: the phenomenological Ginzburg-Landau theory and the microscopic BCS theory.

Generalizations of these theories form the basis for understanding the closely related phenomenon of superfluidity, because they fall into the Lambda transition universality class, but the extent to which similar generalizations can be applied to unconventional superconductors as well is still controversial. The four-dimensional extension of the Ginzburg-Landau theory, the Coleman-Weinberg model, is important in quantum field theory and cosmology.

HISTORY

Superconductivity was discovered in 1911 by Heike Kamerlingh Onnes, who was studying the resistance of solid mercury at cryogenic temperatures using the recently-discovered liquid helium as a refrigerant. At the temperature of 4.2 K, he observed that the resistance abruptly disappeared. In subsequent decades, superconductivity was found in several other materials. In 1913, lead was found to superconduct at 7 K, and in 1941 niobium nitride was found to superconduct at 16 K.

The next important step in understanding superconductivity occurred in 1933, when Meissner and Ochsenfeld discovered that superconductors expelled applied magnetic fields, a phenomenon which has come to be known as the Meissner effect. In 1935, F. and H. London showed that the Meissner effect was a consequence of the minimization of the electromagnetic free energy carried by superconducting current.

In 1950, the phenomenological Ginzburg-Landau theory of superconductivity was devised by Landau and Ginzburg.This theory, which combined Landau's theory of second-order phase transitions with a Schrödinger-like wave equation, had great success in explaining the macroscopic properties of superconductors.

In particular, Abrikosov showed that Ginzburg-Landau

theory predicts the division of superconductors into the two categories now referred to as Type I and Type II. Abrikosov and Ginzburg were awarded the 2003 Nobel Prize for their work.

Also in 1950, Maxwell and Reynolds et al. found that the critical temperature of a superconductor depends on the isotopic mass of the constituent element. This important discovery pointed to the electron-phonon interaction as the microscopic mechanism responsible for superconductivity.

The complete microscopic theory of superconductivity was finally proposed in 1957 by Bardeen, Cooper, and Schrieffer. Independently, the superconductivity phenomenon was explained by Nikolay Bogolyubov. This BCS theory explained the superconducting current as a superfluid of Cooper pairs, pairs of electrons interacting through the exchange of phonons. For this work, the authors were awarded the Nobel Prize in 1972.

The BCS theory was set on a firmer footing in 1958, when Bogoliubov showed that the BCS wavefunction, which had originally been derived from a variational argument, could be obtained using a canonical transformation of the electronic Hamiltonian. In 1959, Lev Gor'kov showed that the BCS theory reduced to the Ginzburg-Landau theory close to the critical temperature.

In 1962, the first commercial superconducting wire, a niobium-titanium alloy, was developed by researchers at Westinghouse. In the same year, Josephson made the important theoretical prediction that a supercurrent can flow between two pieces of superconductor separated by a thin layer of insulator. This phenomenon, now called the Josephson effect, is exploited by superconducting devices such as SQUIDs. It is used in the most accurate available measurements of the magnetic flux quantum,

$$\Phi 0 = \frac{h}{2e}$$

and thus (coupled with the quantum Hall resistivity) for Planck's constant h. Josephson was awarded the Nobel Prize for this work in 1973.

In 2008 it was discovered by Valerii Vinokur and Tatyana Baturina that the same mechanism that produces superconductivity could produce a superinsulator state in some materials, with almost infinite electrical resistance.

HIGH TEMPERATURE SUPERCONDUCTIVITY

Until 1986, physicists had believed that BCS theory forbade superconductivity at temperatures above about 30 K. In that year, Bednorz and Müller discovered superconductivity in a lanthanum-based cuprate perovskite material, which had a transition temperature of 35 K.

It was shortly found by M.K. Wu et al. that replacing the lanthanum with yttrium, i.e. making YBCO, raised the critical temperature to 92 K, which was important because liquid nitrogen could then be used as a refrigerant (at atmospheric pressure, the boiling point of nitrogen is 77 K).

This is important commercially because liquid nitrogen can be produced cheaply on-site with no raw materials, and is not prone to some of the problems (solid air plugs, et cetera) of helium in piping. Many other cuprate superconductors have since been discovered, and the theory of superconductivity in these materials is one of the major outstanding challenges of theoretical condensed matter physics.

From about 1993 the highest temperature superconductor was a ceramic material consisting of thallium, mercury, copper, barium, calcium, and oxygen, with Tc=138 K.

In February 2008 an iron-based family of high temperature superconductors was discovered. Hideo Hosono of the Tokyo Institute of Technology and colleagues found that lanthanum oxygen fluorine iron arsenide (LaO1-xFxFeAs) a oxypnictide superconducts below 26 Kelvin.

Subsequent research from other groups suggests that replacing the lanthanum in LaO1-xFxFeAs with other rare earth elements such as cerium, samarium, neodymium and praseodymium leads to superconductors that work at 52 K. Experts hope that having another family to study will also lead to a theory of the cuprate superconductors.

APPLICATIONS SUPERCONDUCTIVITY

Because of their lack of resistance, superconductors have been used to make electromagnets that generate large magnetic fields with no energy loss. Superconducting magnets have been used in diagnostic medical equipment, studies of materials, and in the construction of powerful particle accelerators.

Using the quantum effects of superconductivity, devices have been developed that measure electric current, voltage, and magnetic field with unprecedented sensitivity. The discovery of better superconducting compounds is a significant step toward a far wider spectrum of applications, including faster computers with larger storage capacities, nuclear fusion reactors in which ionized gas is confined by magnetic fields, magnetic levitation (lifting or suspension) of high-speed ("Maglev") trains, and perhaps most important of all, more efficient generation and transmission of electric power. The 1987 Nobel Prize in physics went to West German physicist J. Georg Bednorz and Swiss physicist K. Alex Müller for their discovery of materials that are superconductive at temperatures higher than had been thought possible.

CONDUCTOR

The maximum pinning force and the maximum current density sustainable by a conductor increases with reduced temperature. Similarly, as the magnetic induction, B, increases, the sustainable current density decreases. Consequently, the current density, magnetic field, and critical temperature are all interdependent.

By increasing any of these parameters to a sufficiently high value, superconductivity can be destroyed and the conductor will revert to a normal, non-superconducting state. One of the most crucial factors in determining the performance of the final magnet is the design of the conductor. This design affects the ultimate field achieved by the magnet, the rate at which the magnet can be energized and the drift rate in the persistent mode of operation. Several phenomena observed in magnets are caused by the conductor itself.

One of the earliest phenomena observed in superconducting magnets wound with single filament conductors was flux jumping. This phenomena arises from current induced in the conductor by the presence of the transverse field generated by the magnet. If a superconductor is placed transverse to the magnetic field, currents are induced in the conductor which shield the bulk of the conductor from the external magnetic field. These circulating currents extend for a finite length along the conductor, flowing in one direction on one side of the conductor and returning on the other side to complete the circuit.

If the heat produced by the penetration of the magnet field into the superconductor cannot escape to the surface rapidly enough, then the temperature increase inside the superconductor triggers a runaway condition known as a flux jump. Consequently, the conductor is driven into the resistive state at fields and currents low in comparison with the critical values. Resistive heat is then dissipated in the small normal zone which increases in temperature causing the normal zone to expand and propagate both along the length of the conductor and transverse to it. This results in the magnet being discharged as the energy in the magnet is dissipated in the resistive portion of the conductor.

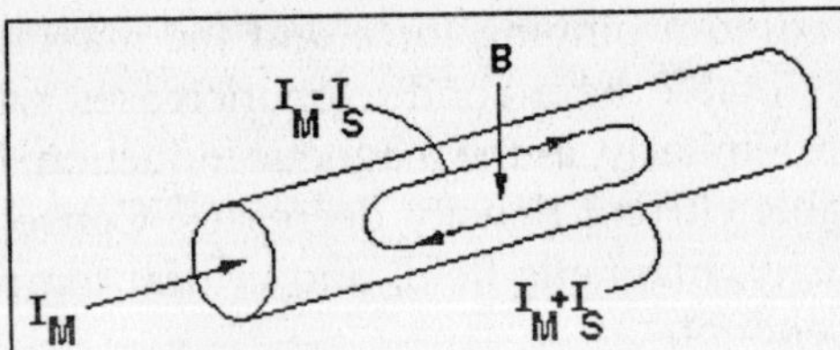

Single Filament Superconductor

To decrease the problem of transition to the normal state, it is common practice to shunt the conductor with a low resistivity normal metal by embedding the superconductor in a copper matrix to form a composite conductor. The copper provides additional heat capacity as well as providing a path for the magnet current while the superconductor is driven normal during a flux jump. If the resistance of the copper is low enough, the temperature of the conductor can remain below the critical temperature at the ambient field, and superconductivity will resume after the currents in the superconductor have decayed.

Embedding the superconductor in a low resistivity metal matrix is effective in reducing the chance of a flux jump which can cause a magnet quench. Magnets constructed with this type of material are dissipative and the heat generated during a flux jump must be conducted to the helium bath. Thus, the magnetic field must be changed slowly to allow time for the heat to be conducted to and dissipated in the liquid helium. Also, the diamagnetic currents in the superconductor contribute to the field generated by the magnet and can reduce its homogeneity.

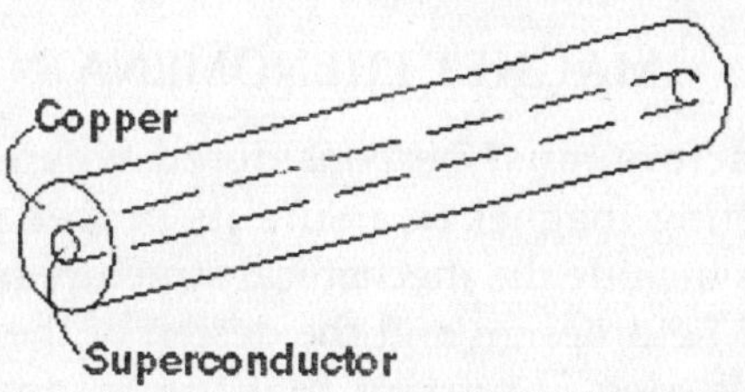

Copper Shunted Superconductor

If, instead of one superconducting filament, many fine filaments of superconductor are used, the heat generated in individual filaments can easily be conducted a short distance to the filament surface thus avoiding flux jumps. Consequently, conductors are made in which many fine filaments of superconductor are coextruded and drawn in a matrix of either copper or aluminum stabilizers. Although this has the desired effect of avoiding flux jumping, circulating currents can again be formed if the conductors are parallel in the highly conductive normal matrix. In this case, the circulating current is between two or more filaments in parallel and the current crosses over through the normally conductive matrix. This gives rise to diamagnetism and unequal distribution of currents in the filaments that limits the rate at which the magnet can be charged.

Problems arising from constructing the superconductor from filaments have been largely circumvented in modern conductors by twisting the filaments in the conductor. This causes the flux from the external magnetic field to be alternated through successive superconducting filaments, thereby reducing the unequal distribution of currents between the superconducting filaments and reducing the diamagnetism of the conductor.

This reduction in diamagnetism or hysteresis has two desirable effects. First, it reduces the amount of energy dissipated in the magnet and permits it to be charged more rapidly. Secondly, the reduced diamagnetism causes the current in the magnet and the magnetic field generated by the magnet to be more linearly related. Such conductors are known as intrinsically stabilized conductors.

MAGNET PHENOMENA

Many factors must be considered when designing a superconducting magnet to assure its proper performance. These factors include the mechanical structure of the magnet, the magnetic field design and the design of the conductor to be used. It is also important that the magnet be able to withstand the mechanical stresses caused by the magnetic and thermally induced forces encountered during normal operation, and the electrical voltages encountered during a quench.

The following paragraphs provide a brief review of some of the more important topics which will assist you in selecting a magnet.

QUENCHING

An important phenomenon in superconducting magnets is quenching. Any superconducting magnet can be quenched by increasing the current and field indiscriminately. A quench in a well encapsulated magnet typically occurs at the location of the highest field in the magnet. Resistance is restored to the conductor at this point and heating occurs in the magnet. This heat spreads to adjacent areas and drives more of the conductor normal, and the normal zone continues to spread until the magnet is completely discharged.

If the resistance across the terminals of the magnet due to the power supply is low, the power supply may be ignored to a first approximation in analyzing a quench. Thus, the quench may be viewed as the discharge of an inductor into a time varying resistance. The resistive voltage, iR, is counteracted by an inductive voltage, Ldi/dt. Unlike the few volts used in charging the magnet, the voltages encountered during a quench discharge can be measured in kilovolts. Initially, the iR voltage is confined to the layers of windings near the point where the quench initiated and internal arcing can occur between layers if sufficient insulation has not been provided.

During a quench a magnet can be damaged by high voltage, high temperature, and high forces. The magnet manufacturer takes all of these issues into consideration as part of the design. Although the magnet is designed to withstand an occasional accidental quench, quenches can shorten the useful life of the device. The photograph above shows an example of a large cold helium gas cloud that forms quickly when a large magnet quenches.

TRAINING

The heat capacities of the materials in a superconducting magnet at 4K are several orders of magnitude lower than the heat capacities of the same materials at room temperature. Thus, a small amount of heat dissipated inside the magnet can raise the temperature of the conductor above its critical temperature at the ambient field and current density.

One source of heating is wire motion caused by the Lorentz force on the conductor in the magnet. Imperceptible motions of the wire can result in frictional heating sufficient to drive the conductor normal at fields well below the anticipated maximum field of the magnet. Upon reenergizing the magnet, it is frequently observed that it will "train" to successively higher fields before quenching, ultimately achieving the design field. In some cases, the wires will remain in their shifted positions and the magnet will perform well. In other cases, however, retraining is required after the magnet is warmed to room temperature. To avoid this training effect, it is necessary that the conductors be securely bonded in place to prevent them from moving.

Bonding the conductor in the magnet entails a substantial risk in that the conductor cannot be recovered and reused after it has been bonded in place.

Since thermal conductivities are very low at these temperatures, the material used to bond the conductor also limits the thermal conductivity of the magnet. Consequently, the effects of wire motion are amplified in that the heat generated is less effectively dissipated to the liquid helium. The conductors in most laboratory sized magnets are bonded with epoxy. Most AMI magnets are wet wound using a filled high thermal conductivity epoxy that is too viscous for vacuum impregnation. Since each turn of the windings is visible as the magnet is being wound, voids in the epoxy can be avoided. Also, the relatively high thermal conductivity of this epoxy causes the heat generated during a quench to be better distributed throughout the coil, thereby reducing the thermal stresses caused by the quench.

Premature quenching can also occur if the large forces between coil sections result in the motion of one coil with respect to another. This is most likely to occur in magnets having coils that are wound in opposition. Such coils are used in bucking coil magnets and magnets for nuclear demagnetization where a low field region is required close to a high field region.

The alternative to bonded windings is to wind the magnet in such a manner that liquid helium permeates the windings. In this case, the liquid helium, which does have a high heat capacity at these temperatures, absorbs the heat generated when the windings move.

The conductor used in this type of construction incorporates a much larger amount of normal conductor to limit the electrical resistance and consequently the temperature increase when the superconductor is driven normal. The operating current density in this type of cryostable magnet is much lower than in the intrinsically stabilized magnets described earlier.

PERSISTENT MODE

After it has been energized, a superconducting magnet can be operated in the persistent mode by short circuiting the magnet with a superconductor. This is accomplished by connecting a section of superconducting wire across the terminals of the magnet. This section of superconductor can be heated to drive it into the resistive state so a voltage can be established across the terminals and the magnet can be charged or discharged.

During the persistent mode of operation, the heater is turned off and the switch is permitted to cool into the superconducting state. In this condition, the power supply may be turned off and the magnet current will circulate through the magnet and the persistent switch.

The decay of the magnet is given by:

$$H = Hoe\text{-}T/t$$

where T is the usual L/R time constant. The small residual resistance in the magnet occurs either from resistance in the joints or from the flux motion resistance discussed earlier.

To achieve the best persistence, the magnet must be operated at less than the maximum field to reduce flux flow resistance in the conductor. High persistence magnets are bulkier, more costly, and require more liquid helium for cooldown than magnets having somewhat less persistence. Nevertheless, these magnets are quite desirable where great persistence is required, such as in nuclear magnetic resonance experiments. In addition, it is necessary that the resistance in joints between conductors be as low as possible.

CHARACTERISTICS OF SUPERCONDUCTING MAGNETS

The most outstanding feature of a superconducting magnet is its ability to support a very high current density with a vanishingly small resistance. This characteristic permits magnets to be constructed that generate intense magnetic fields with little or no electrical power input. This feature also permits steep magnetic field gradients to be generated at fields so intense that the use of ferromagnetic materials for field shaping is limited in effectiveness. Since the current densities are high, superconducting magnet systems are quite compact and occupy only a small amount of laboratory space.

Another feature of superconducting magnets is the stability of the magnetic field in the persistent mode of operation. In the persistent mode of operation, the L/R time constant is extremely long and the magnet can be operated for days or even months at a nearly constant field, a feature of great significance where signal averaging must be performed over an extended period of time.

Small superconducting magnets are frequently used to attain field intensities, stabilities or profiles that are not attainable with alternative magnets — or because their cost is less than the ccst of conventional magnets offering comparable to or inferior performance. In large magnets, the trade-off is frequently made in favour of superconducting magnets based on the relative costs of power for operating the magnets. The cost trade-off becomes more favorable for superconducting magnets as the period of operation increases.

Magnetic field intensities of 1 Tesla or less, without demanding stability requirements, are frequently better generated with water cooled copper coils with or without iron.

Materials and Performance

Most superconducting magnets are wound using conductors which are comprised of many fine filaments of a niobium-titanium (NbTi) alloy embedded in a copper matrix. These conductors have largely replaced the single filament conductors since their magnetic field more readily penetrates the fine filaments, resulting in greater stability and less diamagnetism. Consequently, the linearity of the magnetic field and the magnet current is greatly improved. Another advantage of these conductors is the more rapid rate at which the magnet can be charged and discharged — typically a few minutes for most laboratory size magnets.

Although most magnets are wound with multifilamentary niobium-titanium conductors, some are constructed with multifilamentary, niobium-tin (Nb3Sn) conductors and some with single filaments of niobium-titanium. Nb3Sn conductors are used when the field experienced by the conductor is in excess of about 9 Tesla (90 kilogauss). Typical magnets of this type are wound with a combination of NbTi windings in the low field region and Nb3Sn windings in the high field region. Since multifilamentary Nb3Sn is expensive, brittle, and difficult to wind, these magnets cost more than NbTi magnets.

Single filament NbTi magnets are preferred where the stability of the magnetic field over a long period of time is essential — usually in nuclear magnetic resonance measurements. Better persistent mode operation can be obtained with this material, and since the field is held constant for long periods of time, the extra time required to charge the magnet is inconsequential.

During a quench, the magnet generates high internal voltages and locally elevated temperatures. These cause electrical and mechanical stresses in the windings. The consequences of a quench depend on the design of the magnet and its auxiliary equipment. Permanent damage to the magnet can occur. Normal operation of the magnet at the specified temperature and at magnetic fields equal to or less than the rated field are not expected to cause damage to the magnet, and a warranty covering this type of operation accompanies each magnet.

Niobium-titanium magnets are sometimes operated at temperatures below the normal boiling temperature of liquid helium (4.2K) to achieve even higher fields. Typically, an 8 Tesla solenoid will achieve 9.5-10 Tesla when operated at 2K. AMI magnets are rated in terms of their performance at 4.2K. Fields achievable at lower temperatures are only guaranteed if the magnet is designed for lower temperature operation. Some improvement in performance can also be achieved by reducing the temperature of Nb3Sn magnets, but the increase in field is not as significant as it is in NbTi magnets. When operated at reduced temperatures and higher fields, the energy in the magnet can be increased by 50% or more. Consequently, the magnet might be irreparably damaged if a quench occurs and the magnet is not sufficiently protected. This type of operation should not be attempted without checking with the manufacturer of the magnet – you may invalidate your warranty.

Magnetic Field Intensity

Underspecifying the intensity and homogeneity of the magnetic field used in your experiments may seriously impact your experimental results. However, overspecifying these parameters can greatly increase the costs.

An economic compromise occurs in magnets in which the field experienced by the windings exceeds about 9 Tesla, which is the highest field at which NbTi superconducting alloys can conveniently be used at 4.2K. Higher fields can be attained at this temperature using Nb3Sn conductors, but the increase in cost deserves careful attention. A 10 Tesla magnet operating at 4.2K is substantially more expensive than a 9 Tesla magnet of the same size.

Homogeneity Specifications

Different manufacturers of homogeneous magnets have adopted different standards for specifying the homogeneity of their magnets. Most manufacturers specify the homogeneity in terms of the width of the resonant signal at half the signal height.

AMI uses a more conservative approach that measures the magnetic field at various points in the specified homogeneous volume using a small volume NMR sample. Consequently, small deviations at any point in the volume will be detected. Using an NMR sample equal to the homogeneous volume does not necessarily reveal such small deviations.

If a region of the sample occupying 10% of the specified volume resided in a field having a maximum inhomogeneity five times greater than specified, it is likely that this inhomogeneity would not be noticed. The reason is that the area under this part of the curve is only 10% of the total area and that the resonant line width is five times as broad. Consequently, this inhomogeneous region results in a long shallow tail on the base of the resonant signal.

The cost of homogeneity may be misleading. Homogeneities of ±0.1% in a one-centimeter diameter spherical volume (DSV) are routine. Homogeneities of ±0.001% in the same volume require larger magnets and considerably more effort in their construction. Still more homogeneous magnets require that separately energized superconducting or room temperature trim coils be employed. These coils can escalate the cost quite rapidly. The inside diameter of the magnet scales approximately with the diameter of the homogeneous region specified.

Persistent Switches

Persistent switches are provided on many magnets to increase their stability over long periods of time or to reduce the rate of helium boil-off associated with continually supplying current to the magnet.

A persistent switch is comprised of a short section of superconducting wire connected across the input terminals of a magnet and an integral heater used to drive the wire into the resistive, normal state. When the heater is turned on and the wire is resistive, a voltage can be established across the terminals of the magnet and the magnet can be energized. Once energized, the heater is turned off, the wire becomes superconducting and further changes in the magnet current cannot be made. In this persistent mode of operation, the external power supply can be turned off to reduce the heat input to the helium bath and the current will continue to circulate through the magnet and the persistent switch.

Persistent mode current switches are installed on and become an integral part of the magnet. This is necessary since special care must be taken in making the joints between the switch and the magnet leads. For a typical switch, the electrical heater in the persistent switch has a nominal resistance of 60 Ohms and requires 35 mA of current to drive the superconductor into the resistive state. The superconductive wire typically has 15 to 20 Ohms of resistance in the normal state. AMI's standard persistent switch is 1.25 inches high and has a maximum outside diameter of 1.0 inches.

ELECTRICAL CONDUCTOR

Electrical Conductor, any material that offers little resistance to the flow of an electric current. The difference between a conductor and an insulator, which is a poor conductor of electricity or heat, is one of degree rather than kind, because all substances conduct electricity to some extent. A good conductor of electricity, such as silver or copper, may have a conductivity a billion or more times as great as the conductivity of a good insulator, such as glass or mica. A phenomenon known as superconductivity is observed when certain substances are cooled to a point near absolute zero, at which point their conductivity becomes almost infinite. In solid conductors the electric current is carried by the movement of electrons; in solutions and gases, the electric current is carried by ions.

APPLICATIONS OF TEMPERATURE SUPERCONDUCTORS

The motivation for the development of superconducting coated conductors lies in the many applications for which they could provide superior performance to the materials which are currently in use.

As superconductors are more efficient than conventional materials, significant energy savings in the power generation and distribution industry may be made, with obvious economic and environmental gains.

Also superconducting materials are often necessary for applications which require high current densities, not obtainable using normal metals. The advantages of high Tc materials over low Tc conductors such as Nb3Sn and NbTi include the higher upper critical field, Bc2, and higher critical current density, Jc, as well as the fact that they can be used at higher temperatures leading to potential savings in cryogen costs.

Power Generation and Storage

One potential application of HTS in the power industry is in electrical generators. In addition to the improved efficiency that the use of superconducting rotors would offer, a reduction in size and weight of such a unit is an attractive prospect.

The majority of projects to date have used LTS materials with the most ambitious project undertaken being the Super-GM project in Japan involving the development of a 100 MVA unit. However the prospects for more widespread use of HTS materials in superconducting generators are rather poor partly due to the fact that components presently in use in the electricity generation industry are extremely reliable, but also because of difficulties in reducing AC losses in HTS tapes.

Potential applications in energy storage are superconducting magnetic energy storage (SMES) and superconducting flywheels. SMES works by storing energy in a magnetic field produced by coils which carry a large superconducting d.c. current with no ohmic losses. Such a machine would be useful for meeting diurnal variations in electricity demand, or as an emergency power source.

There are several small units in operation which store around 1 MJ, though these use low Tc materials such as NbTi and significant improvements in the quality of HTS materials are required in order for them to become competitive. Superconducting flywheels use the principle that a superconducting stator may be suspended in mid-air and thus spin with incredibly low friction. Rotational kinetic energy can thus be stored and used to provide power for short periods of time, during temporary faults. Such machines may presently be manufactured using YBCO, but to be of practical use on a large scale, losses will have to be reduced from those presently achievable.

ELECTRICITY TRANSMISSION AND DISTRIBUTION

Potential applications in the electricity distribution industry are more promising. One use of a superconducting wire is to produce a transmission cable – in urban areas, where large amounts of power must be transmitted by underground cables, high current density is important and thus superconducting wires are an attractive proposition. In the US, approximately 100 miles of existing underground cables are replaced each year, representing a market of around $200 million. Since the discovery of HTS materials, several companies have produced cables using BSCCO tapes. One project has recently produced a 30 m long, 3 phase HTS cable which replaced a conventional cable in use at the Southwire company HQ. The cable carries 1250 A at 12.4 kV and has operated continuously for 3000 hours.

There is also interest in the development of transformers, where superconducting materials would have several advantages over those using conventional materials. The main benefit is that losses in the windings are significantly reduced, allowing a reduction in the number of ampere turns used. This leads to a significant reduction in the amount of magnetic flux and hence fewer iron core losses. The main practical advantage is a reduction in the physical size and weight of the coils. Calculations based on an 850 MVA unit have demonstrated that a transformer using HTS materials at 77 K would lead to losses being reduced by a factor of 7 and facilitate a reduction in weight by a factor of 5. A small scale (630 kVA) transformer was manufactured in 1997 using BSCCO and installed in a substation of the Geneva utility company. At present, a 5-10 MVA development, also using BSCCO, aims to supply power to the Waukesha Electric Systems plant in Wisconsin.

Another application of superconducting materials is the fault current limiter, which uses the resistive transition to reduce the fault current in an electrical circuit. Because the transition happens in less than 1 ms, it is possible to prevent damage to circuit components which can be caused by excess current in the time (around 50 ms) taken for conventional circuit breakers to trip. HTS materials are in principle suitable for use in FCLs although most 1st generation conductors are composites containing low resistivity metals (i.e. silver), which would compromise the limiting effect.

Magnet Applications

There are a number of industrial applications based on the ability of superconductors to provide very high field magnets. One such application is in magnetic separation and filtration, which has been performed widely using conventional magnets for some time for example in the enrichment of ferromagnetic iron ores. For small particles which are only weakly ferromagnetic or paramagnetic, much larger fields (and field gradients) are needed for separation. A high profile application of superconducting magnets is to facilitate high speed levitating trains, where the advantages include speed, safety and environmental benefits. A number of projects have been undertaken using conventional magnets such as the German Transrapid system, and test tracks which use LTS materials have also been successfully built, notably the Yamanashi line in Japan. High field magnets are also of importance in research fields such as high energy physics. Particle detector magnets and accelerator beam guidance magnets are both ideally served by superconductors. Notable examples of LTS materials in use at present include the Large Hadron Collider at CERN and the TEVATRON in the US.

Projects using HTS include the development of BSCCO for the 'Very Large Hadron Collider'. A major market for superconducting materials is magnets for magnetic resonance imaging (MRI), primarily for medical diagnostics. As with most other applications, commercial superconducting systems are still based on NbTi. Whilst experimental systems have been built using HTS materials, the main problem is the high cost of BSCCO.

HIGH TEMPERATURE SUPERCONDUCTORS

Until 1986, the record transition temperature for a superconductor was 23 K. In that year, Bednorz and Muller synthesised the compound La2CuO4 which remains superconducting up to 30 K, and soon afterwards other superconducting cuprate materials were discovered with even higher transition temperatures. $YBa_2Cu_3O_7$-d (YBCO) has a Tc of 92 K, which is significant because it is greater than the boiling point of liquid nitrogen at atmospheric pressure. Bi-Sr-Ca-Cu-O, Tl-Ba-Ca-Cu-O and Hg-Ba-Ca-Cu-O compounds have higher critical temperatures.

Table. Transition temperatures of some high T_c superconducting compounds

Material	$T_c(K)$
$YBa_2Cu_3O_{7-\delta}$	92
$Bi_2Sr_2CaCu_2O_8$	85
$Bi_2Sr_2Ca_2Cu_3O_{10}$	110
$TlBa_2Ca_2Cu_3O_9$	123
$HgBa_2Ca_2Cu_3O_8$	135

All these high temperature superconductors have highly anisotropic crystal structures, containing layered CuO_2 planes in which the superconducting charge carriers are thought to be localised.

Chapter 8

Imperfect Solid State

Real crystals are never perfect: they always contain a considerable density of defects and imperfections that affect their physical, chemical, mechanical and electronic properties. The existence of defects also plays an important role in various technological processes and phenomena such as annealing, precipitation, diffusion, sintering, oxidation and others.

It should be noted that defects do not necessarily have adverse effects on the properties of materials. There are many situations in which a judicious control of the types and amounts of imperfections can bring about specific characteristics desired in a system. This can be achieved by proper processing techniques. In fact, "defect engineering" is emerging as an important activity.

All defects and imperfections can be conveniently considered under four main divisions: point defects, line defects or dislocations, planar defects or interfacial or grain boundary defects, and volume defects. We can also add here macroscopic or bulk defects such as pores, cracks and foreign inclusions that are introduced during production and processing of the solid state.

Point defects are inherent to the equilibrium state and thus determined by temperature, pressure and composition of a given system. The presence and concentration of other defects, however, depend on the way the solid was originally formed and subsequently processed.

Briefly consider the effects of imperfections or crystal defects on a few important properties of solids. The electrical behaviour of semiconductors, for example, is largely controlled by crystal imperfections. The conductivity of silicon can thus be altered in type (n or p) and by over eight orders of magnitude through the addition of minute amounts of electrically active dopant elements. In this case, each atom of dopant, substitutionally incorporated, represents a point defect in the silicon lattice.

The fact that such small amounts of impurity atoms can significantly alter the electrical properties of semiconductors is responsible for the development of the transistor and has opened up the entire field of solid state device technology. Practically none of the semiconducting properties that led to these engineering accomplishments are found in a "perfect" crystal. They are properties peculiar to the defective solid state. The existence of dislocations (line defects) in crystals provides a mechanism by which permanent change of shape or mechanical deformation can occur.

A crystalline solid free of dislocations is brittle and practically useless as an engineering material. While the existence of dislocations in crystals insures ductility (ability to deform), the theoretical strength of crystalline solids is drastically reduced by their presence.

We should recognize that dislocations play a central role in the determination of such important properties as strength and ductility. In fact, virtually all mechanical properties of crystalline solids are to a significant extent controlled by the behaviour of line imperfections.

The ability of a ferromagnetic material (such as iron, nickel or iron oxide) to be magnetized and demagnetized depends in large part on the presence of two-dimensional imperfections known as Bloch walls. These interfaces are boundaries between two regions of the crystal which have a different magnetic state.

As magnetization occurs, these defects migrate and by their motion provide the material with a net magnetic moment. Without the existence of Bloch walls all ferromagnetic materials would be permanent magnets. In fact, electromagnets would not exist if it were not for this type of defect.

The presence of surface defects such as cracks causes brittle materials like glass to break at small applied stresses. This fact is familiar to anyone who has broken a glass tube by first filing a small notch (or crack) into the surface.

Removal of cracks from the surface of glass either by etching in hydrofluoric acid or by flame polishing almost always raises the fracture strength.

For example, glass in the absence of any surface cracks has a fracture strength of $\sim 10^{10}$ Newton/m^2 (as opposed to real glass which has a fracture strength of $\sim 10^7$ Newton/m^2).

POINT DEFECTS

Formation of Point Defects

An incontrovertible law of nature states: "Nothing is perfect". This law applies to humans as well as to the inorganic world of crystalline solids and can be formulated as the 2nd law of thermodynamics:

$$F = H - TS$$

where F is the free energy of a given system, H is the heat content or enthalpy and TS is the entropy, or disorder, term. If a reaction takes place at a temperature T, we find the change in F (Δ F) related to a change in H (Δ H), the heat content, and possibly also a change in TS (TΔ S).

Such is the case when defects are formed in a perfect solid: The energy distribution in a solid (Maxwell-Boltzmann) suggests that a number of individual atoms may acquire enough thermal energy to be displaced from the equilibrium lattice site into an interstitial position.

This process of point defect formation requires energy and leads to lattice strain which constitutes, as discussed earlier, an increase in the heat content of the system (Δ H is positive and increases linearly with the number of defects formed). The departure from perfection by the generation of defects leads to disorder (Δ S is positive).

The magnitude of disorder generated (Δ S) is very large during the initial step from perfection to slight disarray, but the increase in disorder (with a given number of defects generated) decreases as the overall disorder increases. Correspondingly the term TΔ S drops rapidly at the beginning and then flattens out.

The net result, free energy, exhibits a minimum for a certain number of defects in the solid [equilibrium defect density = f (temperature)]; the Fminimum suggests also that the transition from perfection to equilibrium defect structure is spontaneous: it occurs naturally.

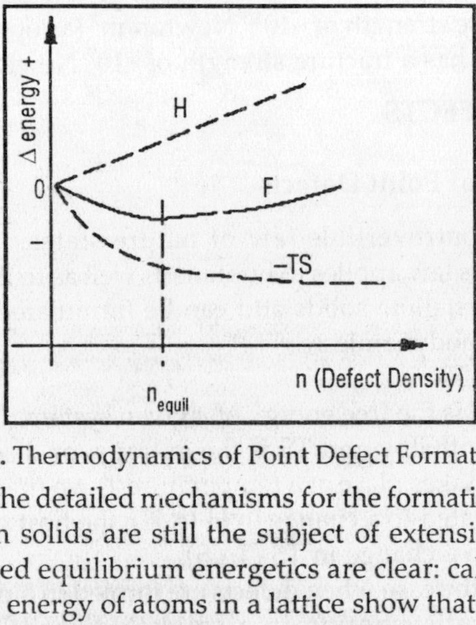

Fig. Thermodynamics of Point Defect Formation.

While the detailed mechanisms for the formation of atomic vacancies in solids are still the subject of extensive research, the associated equilibrium energetics are clear: calculations of the thermal energy of atoms in a lattice show that the average vibrational energy of lattice atoms is much less than 1 eV (the approximate energy change associated with vacancy formation, i.e., the least amount of energy required to form a vacancy) at room temperature.

Therefore a lattice atom will only acquire the energy ΔH_d, the energy required to form the defect, upon the occurrence of a large energy fluctuation. Since the relative probability of an atom having an energy ΔH_d or more in excess of the ground state energy is,

$$e^{-\Delta H_d / kT},$$

the probability that an atomic site is vacant varies in the same way. In a (molar) crystal containing N atomic sites, the number n_d of vacant sites is, therefore,

$$n_d = ANe^{-\Delta H_d / kT}$$

n_d = The number of defects (in equilibrium at T)
N =The total number of atomic sites per mole
DH_d =The energy necessary to form the defect
T = The absolute temperature (K)
k = The Boltzmann constant
A = proportionality constant

Point Defects in "Pure" Metallic Systems

Point defects in "pure" crystalline metals are defects of atomic dimensions, such as impurity atoms, the absence of a matrix atom and/or the presence of a matrix atom in the wrong place.

Some of these point defects. An impurity atom that:

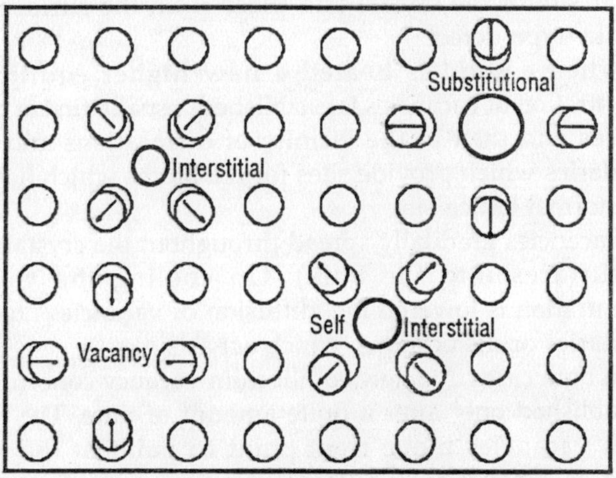

Fig. Point Defects in Crystalline Solids.

Interstitial site depends largely on the size of the atom relative to the size of the site. Small atoms are usually interstitial impurities, while larger atoms are usually substitutional impurities.

A vacancy is an atom site, normally occupied in the perfect

crystal, from which an atom is missing. Often the term "vacancy" is used to denote a so-called Schottky defect, which is formed when an atom or an ion leaves a normal lattice site and repositions itself in a lattice site on the surface of the crystal.

This may be the result of atomic rearrangement in an existing crystal at a high temperature when atomic mobility is high because of increased thermal vibrations.

A vacancy may also originate in the process of crystallization as a result of local disturbances during the growth of new atomic planes on the crystal surface. Vacancies are point defects of a size nearly equal to the size of the original (occupied) site; the energy of the formation of a vacancy is relatively low - usually less than 1 eV.

The number of vacancies at equilibrium at each temperature in a crystal can be determined from equation, in which ΔH_d is the energy necessary to take an atom from a regular site of the crystal and place it on the surface for a Schottky-type defect.

When a solid is heated a new higher equilibrium concentration of vacancies is established, usually first at crystal surfaces and then in the vicinity of dislocations and grain boundaries which provide sites for the atoms which have left their normal lattice site.

Vacancies gradually spread throughout the crystal (from the surfaces into the bulk). On cooling the vacancy concentration is lowered by "diffusion of vacancies" to grain boundaries or dislocations, which act as sinks.

In both cases, the new equilibrium vacancy concentration is established only after a finite amount of time. The rate at which vacancies move from point to point in the lattice decreases exponentially with decreasing temperature.

Thus, on very rapid cooling (quenching) from a high temperature near the melting point most of the vacancies do not have time to diffuse to sinks and are said to be "frozen in". This gives a considerably greater ("non-equilibrium") concentration of vacancies in quenched specimens than that indicated by the thermal equilibrium value.

The concentration of vacant lattice sites in pure materials is very small at low temperatures - about one vacancy every 108 atom sites - and increases with increasing temperature to about one vacancy every 103 sites at the melting temperature.

Vacancies are important because they control the rate of matrix (or substitutional) atom diffusion - i.e., atoms are able to move around in a crystalline solid primarily because of the presence of vacancies.

The mechanism by which they move is the same as that associated with moving a car in a filled parking lot to the exit. Self-interstitials are generally not encountered in close-packed metallic systems, but may be introduced by irradiation.

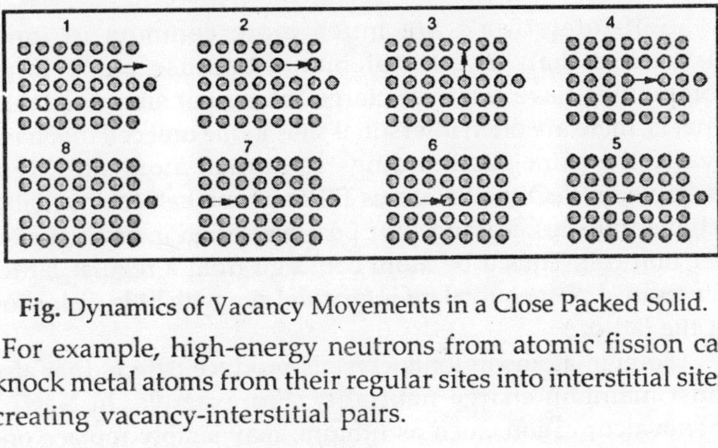

Fig. Dynamics of Vacancy Movements in a Close Packed Solid.

For example, high-energy neutrons from atomic fission can knock metal atoms from their regular sites into interstitial sites, creating vacancy-interstitial pairs.

Point Defects in Ionic Solids

Point defects in ionic structures differ from those found in pure elements because of the charge neutrality requirement. For example, in a pure monovalent ionic material a cation vacancy must have associated with it either a cation interstitial or an anion vacancy to maintain charge neutrality.

Similar requirements hold for anion vacancies. A vacancy pair defect (migration of a cation and an anion to the surface) is usually called a Schottky imperfection, and a vacancy-interstitial pair defect is referred to as a Frenkel imperfection (an anion or cation has left its lattice position, which becomes a vacancy, and has moved to an interstitial position).

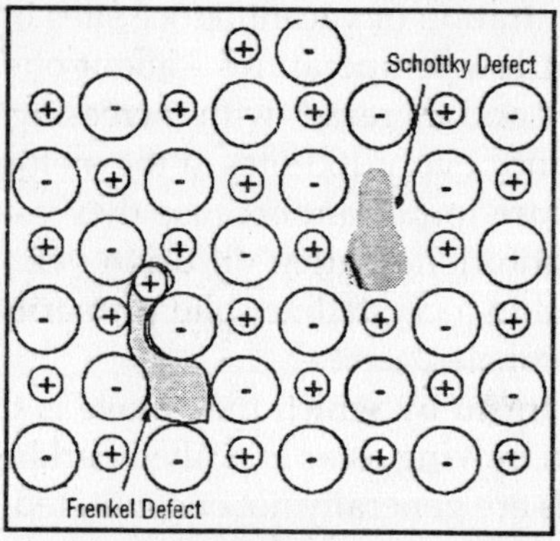

Fig 4 Point Defects in Ionic Solids.

Self-interstitials are much more common in ionic structures than in pure elements because many ionic compounds have relatively large interstitial sites available. That is, there are often interstitial sites in the unit cell that have nearly the same surroundings as normal atom sites. (For example, in BeO the Be atoms fill only one-half the available tetrahedral sites, leaving four possible cation interstitial sites per unit cell. Thus a Be atom could go from a regular lattice site to an almost equivalent interstitial site with little distortion of the lattice.)

Foreign atoms in ionic crystals produce defects that also must maintain charge neutrality. For example, in NaCl a monovalent cation, such as lithium, may simply replace one of the sodium ions as a substitutional impurity. But a divalent cation, such as calcium, replacing a sodium ion must be accompanied by either a cation vacancy or an anion interstitial if charge neutrality is to be maintained. Correspondingly, monovalent impurity cations in a divalent structure (e.g., Na in MgO) must be accompanied by an appropriate number of cation interstitials or anion vacancies.

Point Defects in Covalently Bonded Solids

Substitutional impurities in covalently bonded materials can create a unique imperfection in the electronic structure if the impurity atom is from a group in the periodic table other than the matrix atoms. For example, you already considered Group V and Group III elements in a Group IV matrix, such as As or B in Si.

When foreign atoms are incorporated into a crystal structure, whether in substitutional or interstitial sites, we say that the resulting phase is a solid solution of the matrix material (solvent) and the foreign atoms (solute). The term "solid solution", however, is not restricted to the low solute contents of doped semiconductor systems; there are many solid solutions, such as metallic alloys, that comprise a wide composition range.

LINE DEFECTS

Line imperfections, or dislocations, in crystalline solids are defects that produce lattice distortions centered about a line. A dislocation is simply the edge of an extra inserted fractional plane of atoms. Normally the symbol ⊥ is used to represent a positive dislocation (extra fractional plane) is used to represent a negative dislocation (missing fractional plane).

The importance of dislocations is readily demonstrated in the deformation of crystalline materials. The plane in which a dislocation moves through the lattice is called a slip plane. With an applied shear stress the dislocation moves, atomic row by atomic row, and one part of the crystal is displaced relative to the other.

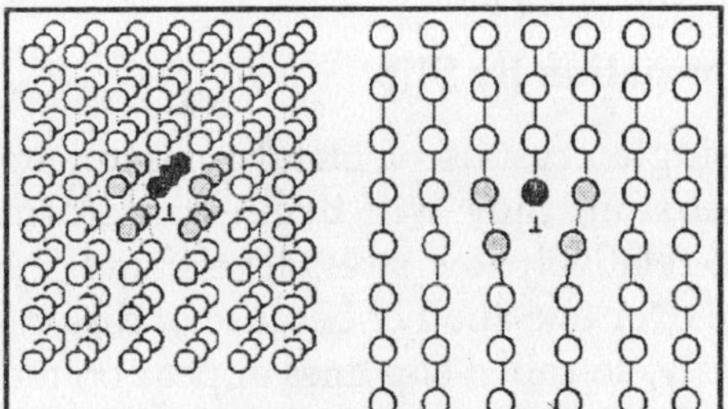

Fig. Schematic Presentation of a Dislocation; The Last Row of Atoms (Dark) in the Inserted Fractional Plane.

When the dislocation has passed through the crystal, the portion of the crystal above the slip plane has shifted one atomic distance relative to the portion below the slip plane. In other words, the motion of the dislocation has caused the crystal to change its shape - to be permanently deformed.

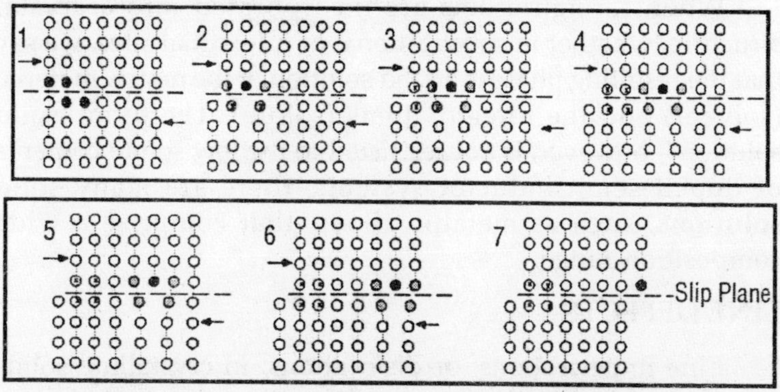

Fig. Plastic Deformation of Crystalline Solid by Slip Associated with Stress Induced Motion of Dislocation.

Please note: on either side of the dislocation the crystal lattice is essentially perfect, but in the immediate vicinity of the dislocation the lattice is severely distorted.

For a positive edge dislocation, the presence of the extra half plane causes the atoms above the slip plane to be put in compression, while those below the slip plane are put in tension. Consequently, the edge dislocation will have a stress field around it that is compressive above the slip plane and tensile below the slip plane.

Plastic Deformation By Slip

When single crystals of metal (or semiconductor) are pulled in tension, they will begin to deform (elongate) plastically at relatively low stress levels, and "blocks" of the crystals slide over one another because of dislocation motion. Simultaneously, so-called slip lines appear on their surface. It is found that deformation by slip occurs most easily on planes with high atomic density and with large interplanar spacing, while the direction of slip is in all instances an atomically "close-packed direction":

For FCC structures we therefore observe as the primary slip system {111} planes in <110> direction, while in BCC structures the primary slip occurs on {110} planes in <111> directions. (It should be noted that an alternate deformation mechanism is "deformation twinning", presently not to be considered.)

Dislocation Climb

Climb is the name given to the motion of dislocations when the extra "half" plane is extended farther into a crystal or partially withdrawn from it. Clearly, the climb process is not a motion of the plane, but rather its growth or shrinking as a result of the addition of atoms or "vacancies" respectively from the environment of the dislocation.

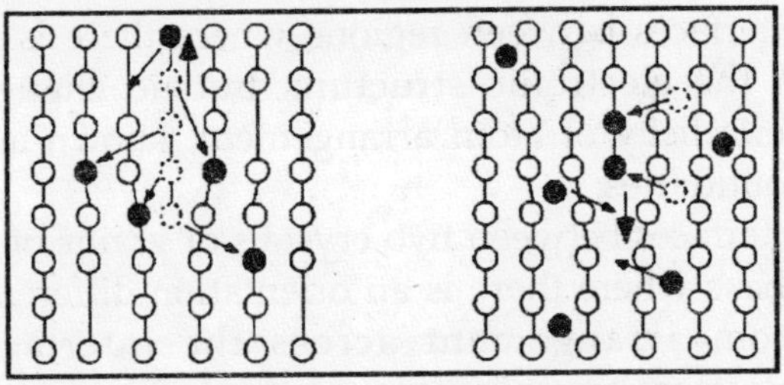

Fig. Dislocation Climb by (a) Loss of Atoms to Surrounding Vacancies and (b) Incorporation of Interstitial Atoms.

Multiplication of Dislocations

Since during slip each dislocation leaves the matrix, macroscopic deformation could not take place given normal dislocation densities in the range of 10^6-$10^8/cm^3$. Examination of the deformed crystals indicates that multiplication of dislocations takes place during deformation.

Dislocation Interactions

The relative ease with which dislocations move across a solid matrix can be attributed to the severe displacements of atoms in the core of dislocations. If these local stresses are reduced, the mobility of dislocations - and thus the ease of slip - is reduced.

It is found that impurities in the vicinity of dislocation cores tend to reduce the local distortion energy of the dislocations and thus stabilize the system against slip. In many systems impurities are intentionally added (e.g., solid solution hardening) to increase the strength of materials. Similarly, micro-precipitates tend to impede dislocation motion (e.g., precipitation hardening).

INTERFACIAL IMPERFECTIONS

The several different types of interfacial, or planar imperfections, in solids can be grouped into the following categories:

- Interfaces between solids and gases, which are called free surfaces;
- Interfaces between regions where there is a change in the electronic structure, but no change in the periodicity of atom arrangement, known as domain boundaries;
- Interfaces between two crystals or grains of the same phase where there is an orientation difference in the atom arrangement across the interface; these interfaces are called grain boundaries;
- Interfaces between different phases, called phase boundaries, where there is generally a change of chemical composition and atom arrangement across the interface.

Grain boundaries are peculiar to crystalline solids, while free surfaces, domain boundaries and phase boundaries are found in both crystalline and amorphous solids.

Free Surfaces

Because of their finite size, all solid materials have free surfaces. The arrangement of atoms at a free surface differs slightly from the interior structure because the surface atoms do not have neighboring atoms on one side. Usually the atoms near the surface have the same crystal structure but a slightly larger lattice parameter than the interior atoms.

Perhaps the most important aspect of free surfaces is the surface energy associated with surfaces of any solid. The source of this surface energy may be seen by considering the surroundings of atoms on the surface and in the interior of a solid. To bring an atom from the interior to the surface, we must either break or distort some bonds - thereby increasing the energy.

The surface energy is defined as the increase in energy per unit area of new surface formed. In crystalline solids, the surface energy depends on the crystallographic orientation of the surface - those surfaces that are planes of densest atomic packing are also the planes of lowest surface energy.

This is because atoms on these surfaces have fewer of their bonds broken or, equivalently, have a larger number of nearest neighbors within the plane of the surface. Typical values of surface energies of solids range from about 10^{-1} to 1 J/m^2. Generally, the stronger the bonding in the crystal, the higher the surface energy.

Surface energies can be reduced by the adsorption of foreign atoms or molecules from the surrounding atmosphere. For example, in mica the surface energy of freshly cleaved material in a vacuum is much higher than the surface energy of the same surface cleaved in air. In this instance, oxygen is adsorbed from the air to partially satisfy the broken bonds at the surface.

Impurity atom adsorption makes it almost impossible to maintain atomically clean surfaces. As a result, surface properties such as electron emission, rates of evaporation and rates of chemical reactions are extremely dependent on the presence of any adsorbed impurities. These properties will be different if the measurements are made under conditions giving different surface adsorption.

Grain Boundaries

Grain boundaries separate regions of different crystallographic orientation. The simplest form of a grain boundary is an interface composed of a parallel array of edge dislocations.

This particular type of boundary is called a tilt boundary because the misorientation is in the form of a simple tilt about an axis, parallel to the dislocations. Tilt boundaries are referred to as low-angle boundaries because the angle of misorientation is generally less than 10°. When a grain boundary has a misorientation greater than 10 or 15, it is no longer practical to think of the boundary as being made up of dislocations because the spacing of the dislocations would be so small that they would lose their individual identity. The grain boundary represents a region a few atomic diameters wide where there is a transition in atomic periodicity between adjacent crystals or grains.

Grain boundaries have an interfacial energy because of the disruption in atomic periodicity in the vicinity of the boundary and the broken bonds that exist across the interface. The interfacial energy of grain boundaries is generally less than that of a free surface because the atoms in a grain boundary are surrounded on all sides by other atoms and have only a few broken or distorted bonds.

Solids with grain boundaries are referred to as polycrystalline, since the structure is composed of many crystals - each with a different crystallographic orientation. In the case of iron the grain boundary structure can be revealed by preferential chemical attack (etching) at the grain boundaries, while the grain structure in polyethylene is revealed by the use of polarized light.

The grain structure is usually specified by giving average grain diameter or by using a scheme developed by the American Society for Testing and Materials (ASTM). In the ASTM procedure the grain size is specified by a "grain size number" (n) where

$$N = 2^{n-1}$$

with N equal to the number of grains per square inch when the sample is viewed at 100X magnification. For example, at a magnification of X = 100, a material with grain size number 8 will show 128 grains per inch2 - this material in effect has (at X = 1) 1.28×10^6 grains per square inch. If the grains are approximately square in cross section, this corresponds to an average grain dimension of 8.8×10^{-4} in.

In polycrystalline samples the individual grains usually have a random crystallographic orientation with respect to one another, and the grain structure is referred to as randomly oriented.

In some instances, however, the grains all have the same orientation to within a few degrees. In this instance the material is said to have a preferred orientation or texture.

Phase Boundaries

A phase is defined as a homogeneous, physically distinct and mechanically separable portion of the material with a given chemical composition and structure. Phases may be substitutional or interstitial solid solutions, ordered alloys or compounds, amorphous substances or even pure elements; a crystalline phase in the solid state may be either polycrystalline or exist as a single crystal.

Solids composed of more than one element may - and often do - consist of a number of phases. For example, a dentist's drill, something painfully familiar to all of us, consists of a mixture of small single crystals of tungsten carbide surrounded by a matrix of cobalt.

Here the cobalt forms a continuous phase. Polyphase materials such as the dentist's drill are generally referred to as composite materials. Composite materials have great importance in the engineering world because they have many attractive properties that set them apart from single-phase materials.

For example, the dentist's drill has good abrasive characteristics (due to the hard carbide particles) and good toughness and impact resistance (due to the continuous cobalt matrix). Neither the tungsten carbide nor the cobalt has both abrasion resistance and impact resistance, yet the proper combination of the two phases yields a composite structure with the desired properties.

The nature of the interface separating various phases is very much like a grain boundary. Boundaries between two phases of different chemical composition and different crystal structure are similar to grain boundaries, while boundaries between different phases with similar crystal structures and crystallographic orientations may be analogous to low-angle grain boundaries in both energy and structure. The concept of a solid consisting of a continuous phase and a discontinuous phase (or phases) leads to a simple classification of the various types of composite materials. Table gives this classification, which is based on the structure (whether amorphous or crystalline) of the continuous and discontinuous phases.

Index

D

E

N

O

P

V

X